B

ITS HISTORY, CULTIVATION, PROPAGATION AND DESCRIPTIONS

By

P. D. LARSON

First Printing Edited by

CHRISTINE A. FLANAGAN

and

CHRISTOPHER F. SACCHI

Second Edition Edited by

JOE METZ

and

JOAN AND SCOT BUTLER

Preface

P. D. "Swede" Larson pursued knowledge of boxwood with a tenacity and enthusiasm shared by few. This work, representing 12 years of effort, was produced out of his practical experience in growing boxwood and educating others about it, his scholarly research, and his sincere belief that boxwood was a much-loved and sorely misunderstood plant. After handling thousands of inquiries about everything from misapplied common names to misguided cultivation practices, he realized that the public hungered for technical information. At the same time, he became convinced that any work on boxwood must also deal with the profusion of cultivars, some of which were beginning to appear in nurseries and public gardens.

Unfortunately, during the final stages of manuscript preparation, Swede succumbed to a long-standing illness. It is with a sense of humility and respect that we have approached the task of bringing his work to posthumous publication.

Swede's knowledge of boxwood and his commitment to caring for the boxwood collection at the Orland E. White Arboretum in Boyce, Va., earned him the title of volunteer curator, a singular honor in the history of our institution. He may be credited with helping to acquire and interpreting one of the largest collections of cultivated boxwood in North America. With the support of the American Boxwood Society, his legacy continues as each section of the arboretum's new Boxwood Memorial Garden, which he helped design, is installed with plants that were propagated by him. Swede contributed to the understanding and promotion of boxwood culture through frequent articles in *The Boxwood Bulletin,* and by serving on the arboretum's board of directors. He was a tireless speaker for garden clubs; gave untold numbers of boxwood garden tours; provided expert instruction during day-long workshops; and propagated hundreds of boxwood cuttings for the arboretum's boxwood collection, plant sales and distribution to plant collections at other gardens. Perhaps most appreciated by the public was his service as horticulturist "on call," to answer inquiries and visit ailing boxwood.

In editing this work, we have added comments only where needed for technical clarity. For consistency, we used the convention that the word "boxwood" may be either singular or plural. In the text, several commercially available products are mentioned by name as examples. No product endorsement is intended or implied.

We are grateful to Jenean Thomson for undertaking the huge job of preparing the figures and illustrations. It has been a pleasure to work with her.

Finally, it is our misfortune that Swede left unfinished his list of acknowledgments. He credits many people who gave generously of their time and knowledge, and whose enthusiasm and encouragement was of great benefit. His partial list included Mary A. Gamble, Burke Davis, William J. Sheehan, Joan Butler and especially Bob Arnold, who gave him space. As for the others, we know you will surely recognize yourselves in between the lines that follow. And, for Bev, one who stood in support of all that he did, we reserve the dedication of this work.

                        Christine A. Flanagan and Christopher F. Sacchi
                        Orland E. White Arboretum
                        Boyce, Virginia
                        June 1, 1996

The Foundation of the State Arboretum of Virginia is grateful to Mrs. P. D. "Bev" Larson for the generous donation of the text copyright and all associated rights. The officers, directors and members of the Foundation offer this book as an enduring legacy to Swede Larson.

BOXWOOD

ITS HISTORY, CULTIVATION, PROPAGATION AND DESCRIPTIONS

Contents

Foreword vi
History, Customs and Legends of Boxwood 1
    A brief look at the history of boxwood
    Boxwood in America
    Customs and legends
Commercial Usage 8
General Characteristics 11
    Species range and distribution
    Natural forms
    Typical leaves
    Flowers and fruits
    Fragrances
    Toxicity and medicinal properties
Cultivation 18
    Soils
    Drainage
    Sunlight
    Exposure
    Watering
    Fertilizers
    Mulch
    Spacing
    Planting and transplanting
    Container-grown plants
    Pruning and renovation
        Maintenance pruning
        Major renovation
    Winter damage – protection and corrective action
    Antitranspirants and antidesiccants
    Most common problems

Propagation 29
   Asexual propagation (cuttings)
   Layering
   Tissue culture
   Sexual propagation
      Seed
      Controlled hybridizing

Pests and Diseases 32
   Boxwood leaf miner
   Boxwood mite
   Boxwood psyllid
   General recommendations
   Boxwood diseases

Nomenclature, Documentation, Registration 37
   Taxonomic nomenclature
   Typographic rules
   Registration

Selected Species, Varieties, Cultivars 41
   Selected Species and Varieties
   Selected Cultivars
   Key to Plant Descriptions
   Plant Descriptions 50
   Plant Descriptions Summary 202

Glossary 216

Selected References 219

Registration List 221

Public Gardens 228

Foreword

The origin of boxwood is ancient. Its history precedes that of humankind. Fossils of boxwood have been found from the Pliocene Age deposits of France (some 1 to 10 million years ago) and some 20 other locations in Europe. Boxwood use in human culture dates from about 4000 B.C. in China and Egypt, and 1000 B.C. in Greece. Through the ages, boxwood has been valued for its wood and as an ornamental plant.

Plants of the genus *Buxus* are indigenous to Central and South America, Europe, Asia and Africa. The earliest American records describe the importation of boxwood, probably from Holland, to Long Island, N.Y., in 1652. As colonists arrived, it spread in favorable climates throughout eastern North America. In later years, European nurseries, especially those in Holland, exported boxwood to North America. However, it was not until the early 1900s that North American plant explorers began bringing boxwood from other areas of the world, especially the Orient, Mediterranean and Middle East.

Little is known about the taxonomic and botanical relationships among the species in the genus *Buxus*. As of 1993, more than 95 species have been discovered in various parts of the world, (Burford, 1993:7) but only about 25 have been thoroughly described in published literature. Over the years, some 60 boxwood cultivars have been placed on the International Registration List of Cultivated *Buxus* L. Additionally, more than 200 cultivars with names are known to be growing in arboreta, commercial nurseries, and private gardens.

Boxwood grows in many natural shapes, sizes and variations — mounded, conical, weeping, spherical, columnar, vase-shaped and billowy pyramidal forms. Sizes range from dwarf to large and tree-like. Leaf colors vary in shades of green and yellow-green, with some having hues of blue and black. A few even show variegations of gold, silver and white. There is a range of leaf shapes and foliage textures as well. Because there are so many varieties of boxwood, it is truly a plant with multiple landscaping possibilities.

Boxwood is a vigorous woody shrub and requires a minimum of care. Only three pests are considered potentially serious: leaf miner, spider mite and psyllid. All are controllable. Diseases are practically non-existent and man-made causes often predispose plants to infection.

This text seeks to present a short story of boxwood's history, cultural care and propagation methods, along with specific details of individual plants. While it is a how-to book aimed primarily at the gardener and boxwood enthusiast, its factual information should prove helpful to horticulturists, growers and landscape architects.

HISTORY, CUSTOMS AND LEGENDS OF BOXWOOD

A Brief Look at the History of Boxwood

The origins of boxwood are indeed ancient and obscure, yet they merit a general chronological narrative to provide perspective.

Pliocene Age.
This period (some 1 to 10 million years ago) has been described as the era when modern plants and animals began their development. Early forms of boxwood have been identified in fossilized Pliocene deposits in France and in some 20 other locations in Europe.

4000 B.C.
The garden of an Egyptian nobleman was described on his tomb in a detailed formal plan. This garden was rectangular in form, with boxwood used to carry out its lines.

2000 B.C.
Chinese gardening started quite early and used flora of unique depth — trees, shrubs (including boxwood) and flowers of almost every family. The gardener's contribution in such circumstances was simply to place these plants in some order or design and emphasize what was already there.

1000 B.C.
In *The Odyssey*, Homer, the great epic poet of Greece, described the garden of Alcinous, where "ships of myrtle sail in seas of Box." In *The Iliad*, Homer mentions that in the Trojan War the yoke of the steeds driven by Priam, king of Troy, was made of box (Iliad, 24:268).

700 B.C.
Tablets made of boxwood are believed to have been used in writing the Bible message of Isaiah. "Now go write it before them in a table, and note it in a book, that it may be for the time to come for ever and ever" (Isaiah, 30:8). Boxwood is mentioned in other references of the period: "I will set in the desert the fir tree, and the pine, and the box tree together" (Isaiah, 41:19); "The glory of Lebanon shall come unto thee, the fir, the pine tree, and the box together, to beautify the place of my sanctuary; and I will make the place of my feet glorious" (Isaiah, 60:13).

590 B.C.

Boxwood is mentioned in the description of the fabled ship of Tyre: "Of the oaks of Bashan have they made thine oars; they have made thy benches of ivory inlaid in boxwood, and from the Isles of Kittum [Cyprus, and other islands of the western Mediterranean]" (Ezekiel, 27:6 revised version). This could very well have been what has been taxonomically classified as *Buxus balearica*, a large tree form of boxwood.

300 B.C.

"The Box tree appears to have first been mentioned by Theophrastus, a Greek horticulturist who lived 372 to 287 B.C. and who ranks the wood with that of ebony, on account of the closeness of its grain" (Loudon, 1844). When the Romans conquered Greece, they copied the Greeks' small boxes (*pyxos*); the Roman name for them was *buxus*.

70-40 B.C.

Virgil called boxwood "smooth grained and proper for the turner's trade which curious hands may carve, and steel with ease invade," and mentioned its use for making musical instruments.

14 B.C.

Vitruvius, celebrated Roman architect, recommended boxwood for topiary work (the Syrians are said to have been the earliest users of topiary) and it was first used by the Romans in the Augustan Age in evergreen sculptures and close-clipped hedges.

23 A.D.

Pliny, the Roman naturalist and writer, left a detailed account of his garden: "...terrace embellished with various figures and bounded with a Box hedge." "...lawn overspread with soft Acanthus surrounded by a walk enclosed with Tree Box, shaped in a variety of forms." "...broad path, laid out in the form of a Circus, ornamented in the middle with Box cut in numberless different figures together with a plantation of shrubs, fenced in by a wall covered by Box rising by different ranges to the top." "...straight walks divided by grass plots, or Box trees cut in a thousand shapes, some forming the Emperor's name, others the name of the gardener."

526 A.D.

One of the first of the Christian monasteries that would rise across Europe during the Middle Ages was built by the Benedictines at Monte Cassino, Italy. In it they copied the architectural concepts of the Roman villa, wherein buildings were arranged around an open peristyle that permitted everyday life to flow readily from indoors to outdoors. Throughout the Middle Ages, the monastery gardens, which were strictly practical and utilitarian, became the reservoir for plants that were essential for culinary, medicinal and cosmetic purposes, as well as house-keeping herbs, fruits and vegetables. Boxwood was on the bare periphery of the medicinal and cosmetic groups, but was sometimes described as a diaphoretic and vermifuge (probably because of its mild toxicity). On the cosmetic side, its leaves yielded a hair dye that produced a fashionable auburn shade.

700 A.D.

The Moorish invaders of Spain used boxwood to create privacy and tall hedges to shield their harem gardens. They also used boxwood and Oriental plant materials to create garden replicas of the Turkish designs seen in their carpets and intricate mosaics.

1300-1500 A.D. (The Renaissance)

Italian pleasure gardens, many of them literally works of art, made elaborate use of boxwood. Italian designers embellished nature and took pride in expressing their natural philosophy and their own point of view. Italian artists and architects, invited to France and Spain to practice their professions and teach, carried this style with variations to other parts of Europe.

The Italian Renaissance gardens, the pattern for other early Renaissance gardens, spawned later variations. The French enlarged the scale of their gardens to include many acres. The Spaniards, who continued the geometric patterns of the Moors, adopted the more highly-ornamented details of Italian gardens. The Dutch, dedicated plantsmen for centuries, became aware of the Italian pleasure garden in the same northern Renaissance as the French. They inherited many of the same stately symmetrical forms, and imitated designs coming out of France. But the needs of prosperous burghers, growing flowers and shrubs in a flat landscape under gray skies, were quite different from those of the aristocratic horticultural showmen of France; a distinctly Dutch style of gardening emerged, characterized above all by its scale. The Dutch gardened in miniature.

The English developed the features of their formal gardens from Italian, French and Dutch styles, adding the long perennial borders for which they are famous. Two types of gardens emerged in early England, those of the Roman legionnaires who constructed the roads and great walls before the birth of Christ, and those which appeared later at the monasteries, cloisters and abbeys, such as Battle Abbey, Hampton Court, Warwick and Hempstead. Unfortunately, Henry VIII, in his 16th-century vendetta against the church, closed the monasteries and destroyed their treasured old gardens. Henry did, however, accept Hampton Court as a gift from Cardinal Wolsey, but Queen Anne — a successor to Henry — disliked boxwood for its smell and ordered the old hedges, arches and borders of Hampton Court cut down. However, by the time of the Renaissance, England, a prosperous and peaceful nation of gardeners, had restored many of the grand old gardens.

Until the 19th century, the northern European gardens of the Renaissance and later periods were primarily planted with forms of one boxwood species, *Buxus sempervirens*, which ranged from England across central and southern Europe, into North Africa and into the borders of Asia Minor.

In Japan, the great Muromachi period of the 14th and 15th centuries created the Gold and Silver Pavilions; the period was followed by a century of bloodshed during which castles and forts of grand style became the order of the day. The 17th century saw a new wave of gardening — the last of the great landscape gardens in which most of the existing lake-and-island gardens were remade. Throughout this entire period, boxwood species, known today as *Buxus microphylla* and *Buxus sinica*, were familiar in the evolving Japanese garden.

1753 A.D.
Carolus Linnaeus, or Carl von Linné (1707-1778), who developed the binomial system of plant nomenclature, assigned the first modern scientific name: *Buxus sempervirens*. He took the generic name *Buxus* from the Latin of the Romans and chose the descriptive name *sempervirens*, or evergreen, for the specific epithet (name) of the species. Linnaeus described two varieties in the landmark book *Species Plantarum* — *Buxus sempervirens* 'Arborescens,' or tree boxwood, and *Buxus sempervirens* 'Suffruticosa,' or little shrub.

Boxwood in America

America's boxwood heritage is the result of a blending of global influences. To begin with, there are no boxwood species known to be native to North America. It was the English gentry who brought many fine specimens of boxwood to their land grants in the mid-Atlantic region of the United States. To the Northeast, Dutch immigrants carried roots of boxwood, and Spanish settlers brought boxwood to the Southeast. The plants were called by such names as English, Dutch, French and Spanish Box.

In later years, European nurseries, especially those in Holland, exported quantities of boxwood to North America as items of commerce. After the American Colonial era, other boxwood species found their way to the United States. *Buxus microphylla*, with an extensive range in Japan, was introduced in 1860. *Buxus balearica*, native to the Mediterranean area, came in the early 1900s. *Buxus harlandii* was first collected by E.H. Wilson in the Hupeh province of China in 1908 and brought to the Arnold Arboretum in Boston, Mass. *Buxus sinica* came from Korea in 1919 under the name *Buxus microphylla* variety *koreana*.

Several individuals and plant societies have been crucial to the study and development of boxwood in America. Dr. Edgar Anderson (1897-1969), a botanist at the Arnold Arboretum and later the Missouri Botanical Garden, introduced open-pollinated boxwood seedlings from the Balkans area in the 1930s. They were to become a significant addition to the horticultural germ plasm of *Buxus sempervirens*.

Henry Hohman (1896-1974), owner of Kingsville Nurseries in Kingsville, Md., was known as one of the best plantsmen this country has produced. He discovered and propagated many fine boxwood cultivars and distributed them to arboreta.

Dr. Henry T. Skinner (1907-1984), curator of the Morris Arboretum and later director of the U.S. National Arboretum, discovered and named several boxwood cultivars and was instrumental in establishing a major boxwood collection while at the U.S. National Arboretum.

Dr. John T. Baldwin, Jr. (1910-1974), cytogeneticist, taxonomist and botanist with the University of Virginia, the University of Michigan, and the College of William and Mary, discovered and named many open-pollinated boxwood seedlings and was an early driving force behind the use of more varied boxwood plants in the landscape.

The American Boxwood Society, founded in 1961 and headquartered at the State Arboretum of Virginia in Boyce, Va., has provided boxwood information to its membership for more than 40 years through its quarterly bulletin. The Boxwood Society of the Midwest, founded in 1976 and headquartered at the Missouri Botanical Garden, has provided boxwood information to its membership for more than 15 years through its biannual bulletin.

There is no doubt that boxwood is deeply rooted in the past. But most Americans know little about storied plants other than impressions gained from public gardens that depict primarily the Colonial era of boxwood. Perpetuation of plant material from that period is the forte of our public gardens. Often, visitors are told that boxwood is American, English, Dutch, French or Spanish boxwood with carefree imprecision. Similarly, terms such as "little leaf," "common," "gold edge," "silver" and "weeping" boxwood do little to help a confused public differentiate among the species and large number of outstanding cultivars now being grown. Even commercial nurseries have contributed to this confusion, although educational efforts of state and national arboreta, as well as those of the boxwood societies, are helping stem the tide.

North America, the land of immigrants, has known virtually every kind of garden, starting with Colonial gardens now restored and maintained at Colonial Williamsburg, Va., the renowned Mount Vernon in Mt. Vernon, Va., and Sylvester Manor in Long Island, N.Y. The progression through modest backyard settings, opulent imitations of French and Italian styles, and serious and imposing versions of English landscape has brought us to a style that is uniquely American, a fusion of influences from almost every corner of the earth. Nowhere is this more evident than in the California style, which can be traced to the 1930s. Seen as the fusion of the house and garden into a single unit, it represents a specifically American synthesis of the spirits of Spanish and Japanese traditions rooted in the distant past. Such movements indicate that 20th-century American horticulture has truly established an independent style.

Although the horticultural endurance and vitality of the genus *Buxus* is apparent, the use of the sturdy evergreen has been remarkably limited. The term "boxwood" is universally known, yet very little is known about it. This is particularly true of the more than 200 boxwood cultivars that have been identified.

With new emphasis on low-maintenance gardening, now is the time for the outstanding boxwood cultivars to come to the fore and, on a new tide of popularity, take their places among the world's most favored garden plants.

Customs and Legends

A number of curious legends, customs and superstitions have been attached to boxwood, as is true for many of our American trees and shrubs.

Boxwood has played a part in religious rites and festivals in many countries through the centuries. In parts of Europe, boxwood was planted around shrines and cemeteries to symbolize immortality. Customs in England's north country used boxwood in funeral ceremonies (Staples, 1971). Mourners in procession carried sprigs of boxwood and threw them into the grave after the coffin had been lowered. In Turkey, widows planted box at their husbands' graves, while the French completely covered new graves with boxwood branches (Dallimore, 1908).

Boxwood was often planted and clipped to form a cross of living green so that no ghost would arise from a grave decorated with boxwood. The ancients used the evergreen leaves of boxwood to symbolize life everlasting for their loved ones who were gone.

Boxwood also figures in legends associated with religious sites and their gardens. Workmen who were employed to build the monastery of St. Christine in the Pyrenees had great difficulty finding a suitable site for a foundation. One morning they saw a white pigeon flying with a cross in its beak. The bird perched on a box tree, and though it flew away on their approach, they found the cross in the branches. Believing this to be a good omen, they built the monastery on the spot where the box tree had stood (Dallimore, 1908).

The origin of boxwood hedges and edges in monastery gardens is traced to the story of Flora, the Roman goddess of flowers, as recalled by the 17th-century Jesuit poet Rapin: "'Gardens of old, nor Art nor Rule obeyed But unadorned, or wild neglect betrayed.' Flora's hair hung undressed, neglected in artless tresses, until in pity another nymph around her head wreathed a Boxen Bough from the fields, which so improved her beauty that trim edgings were placed ever after where flowers disordered once at random grew" (McCarty, 1950:44).

According to an old English Christmas custom, "...unwithering green for garlands hung against the white panelings of lofty rooms. The mistletoe and holly came down at Candlemas and were replaced by boxwood, with its fresh green, to lend cheer until the spring flowers of Easter" (McCarty, 1950:46).

> The box was ordered up,
> Down with the rosemary and bays,
> Down with the mistletoe;
> Instead of holly now upraise
> The greener box for show.
> The holly hitherto did sway,
> Let box now domineer
> Until the dancing Easter day
> On Easter Eve appear.

At one time it was believed that when bees drew nectar from the tiny flowers of boxwood the resulting honey would be spoiled.

To dream of boxwood denoted a long life, prosperity and a happy marriage.

Box was held dear by the maidens of Shakespeare's day: "The leaves and dust of boxwood 'boyld in lye' would make the hair to be of an 'Aboure or Abraham color.'"

From the color of the wood originated the words "buxans pallor" and "buxeous" sometimes used in the sense of spurious, or an allusion to the paleness of the material.

In this country, boxwood was used to hold webs of homespun and flax for benefit of sun and dew, which very well may explain the many fine clumps found in almost inaccessible mountain sections of the Appalachian region.

Thomas Jefferson's Virginia birthplace, Tuckahoe (founded in 1674), is famous for its boxwood maze, box-edged violet beds and the legendary "Ghost

Walk." There, supposedly, a youthful bride sought the shelter of the boxwood to escape from her elderly bridegroom.

Much of the information found in this section has been derived from a 1950 booklet titled *The Story of Boxwood* by Clara S. McCarty, now out of print. McCarty's work was reprinted by the American Boxwood Society in its quarterly *The Boxwood Bulletin*, April and July 1964.

COMMERCIAL USAGE

The wood of box is unique in its combination of fine texture, density, hardness and uniform consistency, yet it readily yields to carving and shaping. Its color, usually a pale off-white, near the shade of old ivory, has made it valuable for inlay.

The use of boxwood for combs can be traced to ancient Egyptians and is common in many countries of the Middle East even to this day.

The ancient Greeks made small chests and boxes called *pyxos*, especially the carved and polished boxes that held fragrant ointments. Through the manufacture of these boxes from the hardwood of a common native shrubby tree, the tree also became known as *pyxos*. To the Romans this tree was *buxus*. And later, in England, it came to be called box.

Among the small objects for which the wood of boxwood has been used and prized over the centuries are pill boxes, both for transporting pills and for individuals to carry. It has been used for marquetry, the inlaying of fine furniture; for chessmen, medallions, religious and secular figurines; for mathematical and musical instruments; and for wood engravings.

Early writers have referred to the use of boxwood. "The turner, engraver, carver, mathematical instrument, comb and pipe makers, give great prices for it by weight as well as measure, and by the seasoning and diverse manners of cutting vigorous insulations, politure and grinding, the roots of this tree... do furnish the inlayers and cabinet-makers with pieces rarely indulated, and full of variety. Also of Box are made wheels or shivers... and pins for blocks and pulleys, pegs for musical instruments, nutcrackers, weaver's shuttles, hollar-sticks, bump-sticks, and dressers for the shoemakers, rulers, rolling pins, pestles, mall-balls, beetles, tops, tables, chess-men, screws male and female, bobbins for bonelace, spoons, nay, the stoutest axle-tree above all" (Evelyn, as quoted in Dallimore, 1908).

Wood engravers and sculptors, both ancient and modern, have found boxwood to be unrivaled as an artistic medium. Although comparable to metal in finish detail, it is superior to it in holding ink. The first wood engravings were books of devotion and playing cards. Sixteenth-century German and Flemish sculptors produced ornate miniature rosary beads, altar pieces and tabernacles. The earliest known boxwood engraving dates from 1423 and portrays St. Christopher carrying the infant Saviour.

Buxus balearica, indigenous to the Mediterranean area, was the boxwood of the early timber trade, valued for the quality and color of its wood, which gleams like old ivory. Its fine, smooth grain and resistance to warping also made it ideal for arts and crafts throughout history. The European industrial revolution created an insatiable demand for reliable wooden shuttles for the looms of England's burgeoning textile industry. The supply of *Buxus balearica* was soon exhausted. Entire natural plantations of these boxwood were denuded of trees described as growing 50 feet tall. Boxwood timber was in such demand during the period of 1860 to 1880 that English imports from the Caucasus, Asia Minor and Persia averaged about 6,000 tons annually (McCarty, 1950).

Loudon, a British horticulturist, wrote in 1844: "The wood of trunks in France is rarely found of sufficient size for blocks, and where it is, it is so dear that trees are not cut down at once, but pieces are taken from living trees as required."

Dallimore reported that during this time "French turners were wont to place the wood in dark cellars for a period of three to five years, to keep it from splitting. It was then taken out, the bark removed, and buried in the ground to keep it from light until required for use." He also stated that in 1815 trees on Box-hill [England], to the cost of £10,000, were cut down.

Then came a sharp decline, hastened by the action of the Russian government. It reportedly offered to sell to a Liverpool firm all the boxwood in the Caucasus for a lump sum of 10,000,000 rubles — with the condition that the purchaser was to build a specified number of military roads. However, after Britain declined the proposal, the Russians imposed an export duty on all boxwood and walnut shipped from the Caucasus. The shuttle-makers soon turned to other, cheaper woods, among them Virginia's flowering dogwood and persimmon (McCarty, 1950).

In the Jacobean era, boxwood was used for the geometrical designs of Indian and Persian motifs, for lines in combination with ebony, and for marquetry during the Queen Anne and William and Mary periods.

Many cabinet makers used boxwood for inlays, false teeth, and ornamentation on cornices of Sheraton-style furniture in combination with mahogany and rosewood.

Boxwood has even contributed to the technology of measurements. Prior to the early 19th century, linear standards in Europe and North America were not established and measuring was a highly individual matter. Each craftsman made his own rule and an inch equalled three round dry barley grains. However, standards of measurement enabled the mass production of folding rules. For rule bodies, most North American companies during the 1800s favored boxwood, a dense, stable wood first imported from Turkey, Russia, and later from Venezuela. By the early 1900s, the folding rule lost its appeal to the more easily extended zigzag rule and the compact coiled-spring tape measure. By 1945, Stanley sold only a few boxwood rules. Today, these early boxwood rules have become collectible antiques.

In Taiwan, the artists who work in ivory do their apprentice work using boxwood. In Japan, "netsuke", those imaginative creatures that serve as buttons on the sashes of men's kimonos, frequently are carved from boxwood.

Although rising in popularity, the harvest of branches by breaking or clipping boxwood is not a new thing in North America — it's been practiced for many years. The clippings are used by florists for ropes, laurels, wreaths, backing and filler.

This short list is only a beginning, but its diversity stirs the imagination with thoughts of how long, close and intimate has been man's association with boxwood.

GENERAL CHARACTERISTICS

Boxwood is a broad-leaved evergreen shrub whose stiff leaves are entire, the margins continuous. The leaves are opposite, and more than one leaf shape and size may occur on a single plant. The internodal length varies slightly, and the upper leaf surface is most often darker than the underside. Most forms have two annual growth periods. The first begins in early spring and the second in late summer or early fall.

Boxwood grows in many natural shapes, sizes and variations. Sizes and shapes range from dwarf shrubs to large, upright, tree-like, or even weeping forms. Leaf colors vary in shades of green and yellow-green, with some having hues of blue and black; others have variegations of gold, silver and white. All of these varying characteristics make boxwood a plant with a wide range of uses.

Figure 1. Worldwide distribution of indigenous *Buxus* species.

Species Range and Distribution

Little is known about the botanical and taxonomic relationships among the species in the genus *Buxus*. However, a consensus of current taxonomic data suggests that there are about 95 to 100 species. Of these, 25 to 30 species have been reasonably well-described; 60 to 70 are lesser known, but have been identified and described to some degree. Species of *Buxus* are distributed among five major geographical regions of the New and Old World, with only temperate North America and Australia lacking any known indigenous species of boxwood.

As shown in Fig. 1, the principal regions for indigenous boxwood include:

1. **Europe, Mediterranean Basin and Middle East**

   Best Known:
   B. *balearica*
   (Spanish Boxwood)
   B. *sempervirens*
   (Common Boxwood)

   Lesser Known:
   B. *longifolia*

2. **China, Japan, Korea, Malaysia and the Philippines**

   Best Known:
   B. *austro-yunnanensis*
   B. *bodineri*
   B. *cephalantha*
   B. *hainanensis*
   B. *harlandii* (Harlands Boxwood)
   B. *hebecarpa*
   B. *henryi*
   B. *ichangensis*
   B. *latistyla*
   B. *linearifolia*
   B. *megistophylla*
   B. *microphylla* (Littleleaf Boxwood)
   B. *microphylla* v. *japonica*
   (Japanese Littleleaf Boxwood)
   B. *mollicula*
   B. *myrica*
   B. *pubiramea*
   B. *rugulosa*
   B. *sinica*
   B. *sinica* v. *insularis*
   (Chinese Littleleaf Boxwood)
   B. *stenophylla*

   Lesser Known:
   B. *cochinchinensis*
   B. *fortunei*
   B. *holttumiana*
   B. *intermedia*
   B. *liukiuensis*
   B. *loheri*
   B. *malayana*
   B. *nepalense*
   B. *pachyphylla*
   B. *pubifolia*
   B. *rivularis*
   B. *rolfei*
   B. *rupicola*

3. **Africa (tropical and southern)**
   Best Known:
   B. benguellensis   B. nyasica
   B. hildebrandtii
   Lesser Known:
   B. calophylla   B. macowani
   B. hirta   B. pedicellata
   B. madagascarica

4. **Caribbean Islands, Mexico and South America**
   Best Known:
   B. bahamensis   B. bartletti
   Lesser Known:
   B. acuminata   B. macrophylla
   B. aneura   B. marginalis
   B. arborea   B. mexicana
   B. brevipes   B. muelleriana
   B. citrifolia   B. obovata
   B. conzattii   B. olivacea
   B. crassifolia   B. pilosula
   B. cubana   B. pubescens
   B. ekmanii   B. pulchella
   B. excisa   B. purdieana
   B. flaviramea   B. retusa
   B. glomerata   B. revoluta
   B. gonoclada   B. rheedioides
   B. heterophylla   B. rotundifolia
   B. imbricata   B. shaferi
   B. laevigata   B. vaccinioides
   B. lancifolia   B. wrightii

5. **India, N.W. Himalayas and the former Soviet Union**
   Best Known:
   B. colchica sempervirens B. wallichiana
   B. himalayensis   (Wallichian Boxwood)
   B. hyrcana sempervirens
   Lesser Known:
   B. papillosa   B. subcolumnaris
   B. pulchella   B. vahlii

Most cultivated boxwood is derived from and has been assigned to the species *Buxus sempervirens*, *Buxus microphylla*, *Buxus harlandii* and *Buxus sinica*. These are all rather widely spread through temperate Asia, Europe and North America.

General Characteristics

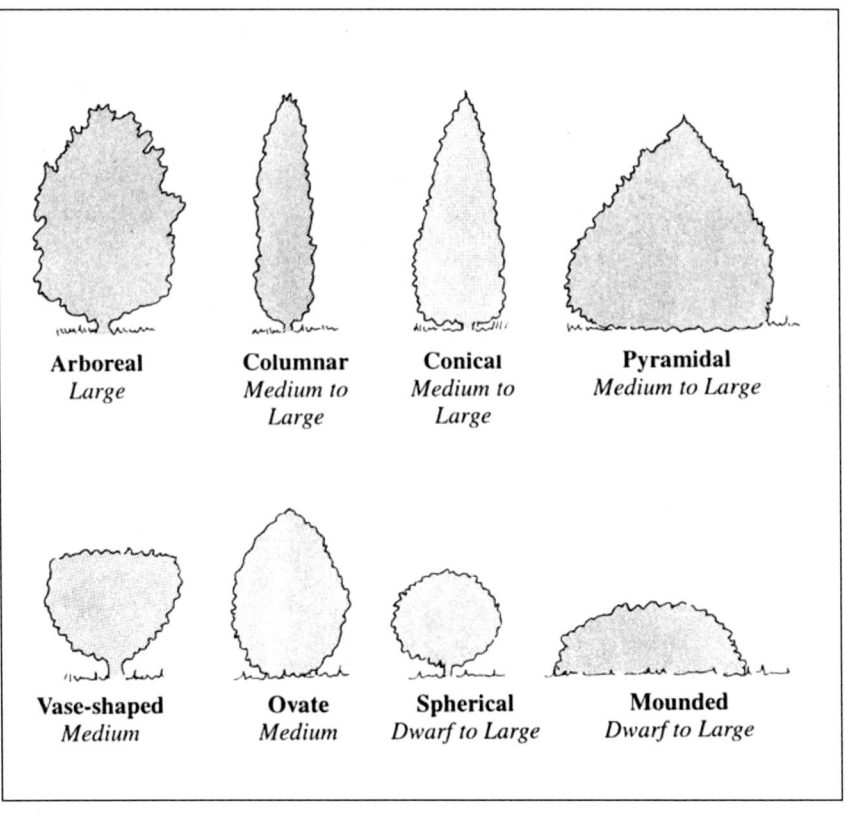

Figure 2. Plant height at 25 years of age. Dwarf: up to $2^{1}/_{2}$ ft., Small: $2^{1}/_{2}$ to $3^{1}/_{2}$ ft., Medium: $3^{1}/_{2}$ to 6 ft., Large: Over 6 ft.

Natural Forms

Boxwood occurs naturally in many shapes, sizes and variations. Sizes range from dwarf to large and even tree-like. Although boxwood typically grows slowly, the variations among species, varieties and cultivars is evident by 10 to 15 years of age and well-established by age 25. Most boxwood exhibits a change in growth characteristics, usually at about 25 to 30 years of age. At this time, the annual growth rate, either in height, width or both, slows and in some cases virtually stops. Consequently, change in the habit or shape, although subtle, may continue to occur; the overall form may change from quite compact, billowy and pyramidal to slightly more open and conical in appearance.

Boxwood have many natural forms that may be categorized into classes (Fig. 2), each with its own range of sizes. The forms defined here are based on extensive studies of identified, living specimens observed at private estates, arboreta and botanical gardens throughout the United States. Much of the work is based on the large collection of boxwood at the State Arboretum of Virginia, in Boyce, Va.

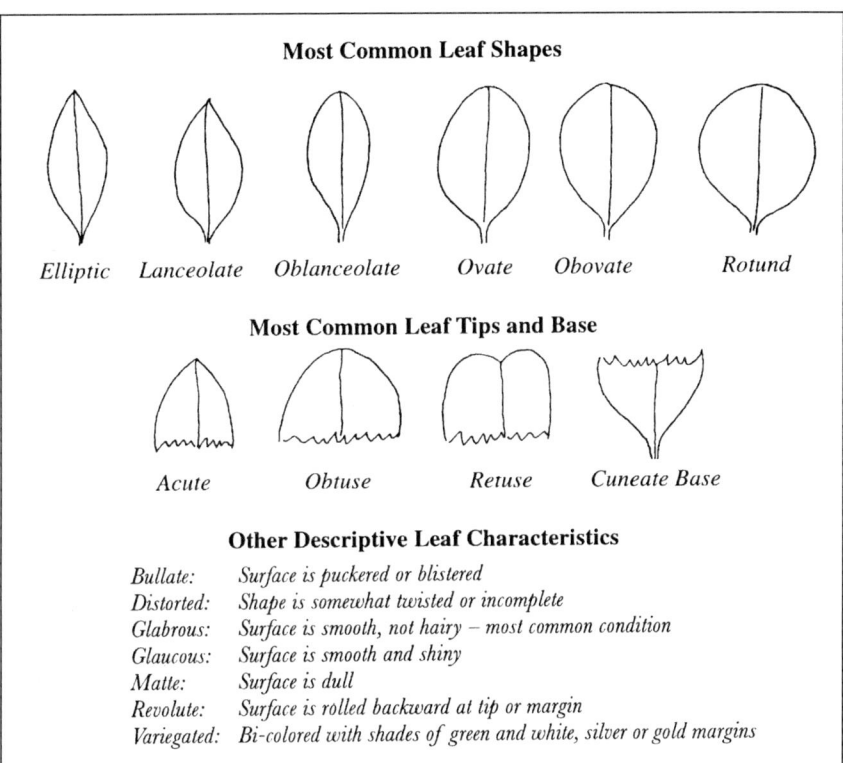

Figure 3. Leaf Characteristics.

Typical Leaves

The following descriptions are paraphrased and adapted from Rehder's *Manual of Cultivated Trees and Shrubs* (Macmillan, 1940) and L. H. Bailey's *Manual of Cultivated Plants* (Macmillan, 1949).

The leaf shapes within the genus *Buxus* vary quite widely, and there is also variation within each species. Typically, leaf shape alone is not sufficient to identify the many existing cultivars. Leaf shape across cultivars of the same species is widely variable; and although one shape is usually predominant, a cultivar may exhibit more than one leaf shape on an individual plant.

Many botanical terms exist for precisely describing leaves. However, only those terms that refer to the leaf shapes, tips and bases most commonly observed in ornamental boxwood have been included here. This set of leaf characteristics should allow a reasonably precise description, particularly when coupled with the plant's natural form, to identify most boxwood.

The most common leaf shapes are depicted in Fig. 3.

The leaf colors of boxwood define a subtle spectrum of greens that range from light through medium to dark green. Some species and cultivars have undertones of yellow-green, others exhibit hues of blue and even black. Some cultivars of boxwood have leaves of varying greens edged, striped, and splotched with white, silver or gold.

General Characteristics

Flowers and Fruits

We tend to overlook boxwood flowers since they have no petals and are rather inconspicuous. However, some cultivars flower profusely and even have a pervasive fragrance. Curiously, not every boxwood cultivar blooms. The late Dr. John T. Baldwin, Jr., a world authority on *Buxus*, noted that some boxwood do not flower and hence produce no fruits. Some cultivars may grow for 15 years or more before they bloom and bear fruit, and some set fruit in which a portion or all the seeds abort. For boxwood that do bloom, the flowers tend to appear in the spring within dates strongly influenced by climatic conditions. Warm temperatures tend to induce early flowering.

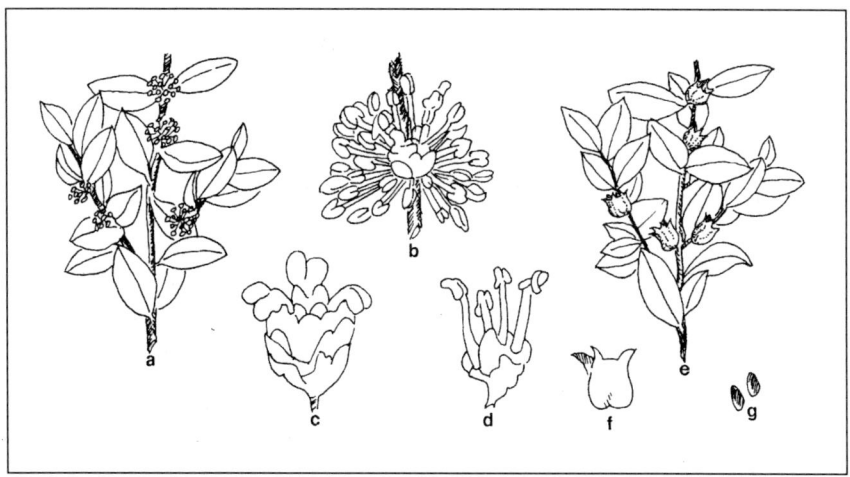

Figure 4. Typical flowers and fruits of *Buxus* species. a: flowering branch, b: flower, c: pistillate flower, d: staminate flower, e: fruiting branch, f: fruit, g: seed. Illustrations are not to scale.

Boxwood flowers (Fig. 4) are either male or female. They occur in axillary or terminal clusters that usually consist of a central pistillate flower (female, Figure 4 c.) and several staminate flowers (male, Figure 4 d.). The pistillate flower has six sepals and a three-celled ovary with three short styles (Figure 4 c.). The staminate flower has four sepals and four stamens, with the stamens much longer than the sepals (Figure 4 d.). The pistillate flowers develop fruits (Figure 4 f.). The fruit is a subglobose, three-horned capsule, which at maturity opens into three two-horned valves; in each valve are usually two lustrous black seeds (Figure 4 g.).

When ripe, the fruits dehisce (open) and seeds may be ejected several feet from the plant. If the soil surface conditions are satisfactory, the seeds will germinate and produce seedlings. The seeds may also be collected and propagated.

Unless careful intervention has been taken to cover unopened flowers and hand-pollinate them, all seeds produced naturally will result from cross-pollination by bees and other insects, i.e., they will be of unknown parentage (at least regarding the male) and thus are not identical.

Fragrances

There are two fragrances that emanate from some boxwood cultivars, one from the leaves and another from the blossoms. Not all boxwood cultivars emit a distinctive fragrance from their leaves, and not all cultivars blossom.

Actually, the entire genus is unjustly blamed for the highly pungent leaf fragrance emanated by the single (and widely used) cultivar *B. sempervirens* 'Suffruticosa,' commonly known as English Boxwood. For many other cultivars, an extremely sensitive nose is required to even detect an aroma.

The fragrance of boxwood has been described by some as foxy due to its slight acridity. Others have called it clearly aromatic. This aroma may be almost absent on a hot, dry day, whereas on a rainy, warm day the odor can become pervasive. Likewise, the more subtle and spicy cinnamon-like aroma of the flowers can be quite penetrating and delightful. Interestingly, the fragrance of boxwood rarely fails to elicit opinions from garden visitors. Katherine Hepburn once said she did not use anything that smells good, such as perfume or scented soap; but she did like the way the boxwood smelled on her grandfather's farm in Virginia, mixed in with the fresh smell of live chickens. Dolley Madison preferred, above all else in her garden, the blending fragrance of roses and the pungency of box. Others have described the boxwood fragrance as cat-like, to put it delicately; but Oliver Wendell Holmes waxed more rhetorical when he called it "the fragrance of eternity." And for those who don't seem to smell boxwood at all, pity or envy as your opinion guides you.

Toxicity and Medicinal Properties

Fear and superstition have been the background for information about the toxic and/or medicinal qualities of *Buxus*. Dodoens, a Dutch physician and botanist of the 17th century, warned of its possible harm, especially to the brain.

Walter Conrad Meunscher (1975) wrote that livestock — cattle, horses, sheep and pigs — browse on boxwood with fatal results, that the plant is emetic and purgative in its action and may cause nervous symptoms and convulsions.

The leaves and stems of at least one species (*B. sempervirens*) do contain an alkaloid called buxine, which makes them slightly poisonous if ingested by humans. Their bitter acrid taste does appear to discourage some animals from nibbling. However, dairy cattle have been observed to browse a form of *B. sempervirens* without apparent distress.

Most reports indicate humans are susceptible, although rarely, to dermatitis from contact with the leaves and stems of boxwood. In fact, various parts of the boxwood plant have been used as cures for rheumatism, malarial fevers and epilepsy.

CULTIVATION

Not all boxwood species, varieties, and cultivars respond to culture and care in precisely the same manner. I will first deal with the general requirements of boxwood. More specific information is contained in the pages listing selected species, varieties, and cultivars beginning on page 41.

Boxwood is a vigorous and long-lived shrub with many specimens known to be more than 200 years old. Most boxwood problems are man-made. In a fairly good environment, boxwood mind themselves, as long as reasonably good cultural practices are observed and they are not seriously attacked by leaf miner, mites or psyllid.

Boxwood grows well in many different soils, and will thrive when sufficient humus is present and the soil is friable. Many yards and gardens filled with subsoils from basements suffer from having been graded when the soil was muddy. Such soil typically bakes in dry weather and stays soggy in wet seasons. Under such conditions, boxwood is not likely to prosper unless drainage is provided in the bottom or from the side of the planting hole.

Soils

Boxwood does not require any specific soil type, but does prefer a well-drained soil. Ideally, soils with an even mix of silt, sand, clay and incorporated organic material provide the best planting medium. Sites with heavy clay or light sandy soil should be avoided or amended.

Soil pH (potential hydrogen) is a numerical measure of the acidity or alkalinity of the water in soil. The pH scale runs from 0, which is extremely acidic, to 14, which is extremely alkaline. The center of the scale, pH 7, is neutral. Most soils of the United States have a pH ranging between 5 and 9. In areas of abundant rainfall, including most of the eastern United States, soils tend to be acidic, since water moving through the soil leaches out alkaline compounds. In the arid regions, including the western United States, soils tend to be alkaline.

Boxwood will grow in soils ranging in pH from 6.5 (slightly acidic) to 7.2 (mildly alkaline) but, if given their choice, prefer a slightly alkaline or "sweet" soil with a pH above 7.0.

There are several ways to measure the pH of your soil. Send a sample to the laboratory operated by the state extension service; or test a sample yourself with a test kit, a pH meter or litmus paper. If your soil is professionally tested, you will also learn about its organic and mineral composition.

Adjustments to the soil can be made with the application of certain material. To make soils *less* acidic, apply a material that contains some form of lime (ground agricultural limestone is the most frequently used). To make soils *more* acidic, apply elemental sulphur or aluminum sulphate. Be sure to follow the directions on the package or the recommendations of your county extension agent.

The texture of the soil, organic content and soil type are all factors to consider in making adjustments to soil pH.

Drainage

Boxwood do not like "wet feet." Their roots require air in the soil to grow well. Air is introduced when the soil is well-drained and water does not stand long after a rain or watering session. When the roots experience water-logged soils, the plant stops growing, slows its intake of water and nutrients, foliage becomes yellow, growth is stunted, and susceptibility to disease is enhanced.

Consider raised beds or installation of a submerged French drain, drain tiles, hollow blocks or similar materials if your soil drains so poorly that simple additions of gravel or sand well-mixed throughout the root zone are not feasible.

Sunlight

Boxwood tolerate shade but grow more rapidly in sites that receive direct sunlight for at least part of the day. Some cultivars, particularly those of the species *B. sempervirens*, do well in full sunlight. A few cultivars sunburn or bronze in full sun, particularly those of the species *B. microphylla* and *B. sinica*. Although most cultivars seem to thrive best with daily exposure to moderate amounts of direct sun, a good general rule to follow is high open shade and morning sun.

Exposure

Boxwood should be placed out of the prevailing winter winds, or at least be protected from them. This is especially true when seasonal extremes occur either in late fall or early spring, and where the extreme conditions are exacerbated by strong winds and plant desiccation.

Sites with significant wind exposure can be modified with mounds of earth (berms), walls, fences or lines of evergreen trees as windbreaks.

Siting boxwood against a reflecting surface with eastern exposure (e.g., a building or smooth wall), invites winter damage from glare induced by snow and ice. Western exposure with persistent cold winds and late afternoon sun is also an invitation to winter damage. In both cases, wide temperature variation is the prime culprit.

Watering

The basic rule is water seldom and thoroughly. Supplement natural rainfall to provide about one inch of water per week. Well-drained soil is essential.

Newly-planted or transplanted specimens require slightly more water for the first year until they become established. Thorough watering at longer intervals is always much more beneficial than frequent light watering or sprinkling. Thorough watering, which moistens the entire root ball and surrounding feeder roots, encourages development of a well-branched, healthy root system.

If you are unable to provide water relief during unusually hot and/or dry periods, avoid actions that stimulate growth or further stress. Don't prune, fertilize, apply pesticides or perform any other cultural practice that will induce growth that a plant under stress will be hard put to support. Just leave it alone — most boxwood cultivars adjust to water stress quite well.

During a spring drought, watering will stave off desiccation as plants break dormancy and growth resumes. When fall drought occurs, make sure your plants enter the dormant season with adequate soil moisture.

Fertilizers

Boxwood should be allowed to follow a natural course of steady even growth for best long-range results. The ideal growth rate often cannot be realized if the plant is stimulated by synthetic fertilizer because it causes uneven surges of rapid growth. I believe that synthetic fertilizers do little to improve precious soil and act only as a quick fix, producing rapid, unnatural and undesirable growth. I recommend dressing with compost, which yields the same priceless natural nutrients that Mother Nature provides for the flora of the forests and field without the help of mankind. This natural way of feeding plants and improving the soil generally produces stronger, healthier plants that are more pest- and disease-resistant.

However, if you believe synthetic fertilizers are a must for your boxwood, here are a few admonitions and suggestions:

- A soil test for your garden is a **must**.
- Fertilizer is wasted unless the correct soil pH (6.5–7.2) is maintained to allow roots to absorb the needed nutrients.
- Fertilize in the late fall.
- Don't fertilize a plant in the first year it is planted or transplanted.
- Fertilize every third or fourth year, if a soil test shows need for nutrients.
- When using synthetic fertilizer, carefully follow the directions that accompany it.
- Boxwood will grow well without fertilizer.

Mulch

A good, properly-applied mulch serves many functions: It facilitates the penetration of water into the soil; conserves and stabilizes moisture; helps prevent soil erosion, and even the baking and cracking of some soils; helps reduce weed growth; reduces extremes in soil temperature; and adds nutrients to the soil.

Use organic mulch rather than inorganic because it further benefits soil texture. Non-acidic and mildly acidic mulches are more desirable for use with boxwood. However, if only the more acidic mulches are available, the addition of ground limestone will help maintain a more neutral pH (see Soils, p. 18). Virtually all organic mulches benefit from aging at least a year before application. Some preferred mulches are shredded hardwood bark; composted wood chips; well-rotted sawdust; pine bark, nuggets and needles; shredded leaves and grass clippings; shredded salt hay, marsh hay and straw; and cocoa bean and buckwheat hulls. Most any organic material that has been composted could be used.

Boxwood do not like a heavy layer of any type of mulch. Apply no more than one inch for established plants, and slightly more for newly set out plants and transplants, particularly while they are becoming established or need winter protection.

Spacing

All boxwood do not have the same space requirements, nor are they always used for the same horticultural purposes. It is clear that spacing is different for single specimens, foundation plantings, groupings for background and area separations, edgings, hedgings, allées, topiary and espalier. It is therefore difficult to produce a general set of numbers.

When planning and planting, first consider the ultimate space requirements and desired long-term landscaping effect, tempered by knowledge of local climatic conditions. Because not all boxwood are the same, the selection of the appropriate boxwood species, variety or cultivar to suit these objectives and conditions is important. The information provided in the plant descriptions and summary should be consulted early in the planning process.

Planting and Transplanting

Carefully prepare the site, taking note of the soil characteristics and providing ample drainage for the roots. Dig a hole as deep as and 6 to 8 inches wider than the root ball of the plant.

If the soil is reasonably friable, use it. If not, replace it with a mix of one part rotted manure or compost, one part loamy soil, and one part coarse sand and peat to create a better planting bed. Improving the soil now will pay off later.

When the hole has been dug and the soil prepared, set the plant on firm ground, slightly higher than its original depth. Fill in around the sides and tamp gently. Leave the ground level from the base of the plant to just beyond the root system or drip line (whichever is greater). Create a slight mound at the outer perimeter. Water slowly and deeply after planting, taking care not to wash soil from over the roots.

Transplanting methods for boxwood will depend entirely on the individual plant: age, size and condition. Young and small plants can be transplanted with less risk than older and larger plants. Healthy plants are more likely to survive the shock than ones under stress.

There is a best season for transplanting. It is the period with the highest and most dependable average rainfall. Most areas in the United States experience this in the spring and/or early fall. Be in rhythm with the season.

As shown in Fig. 5, before you start to dig, go out at least 6 to 8 inches from

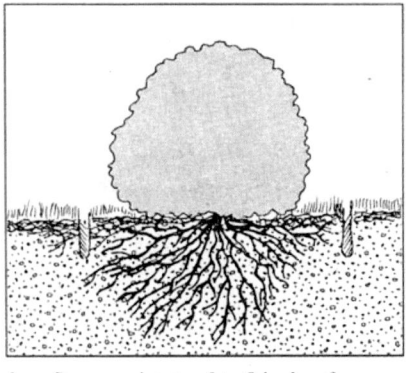

1 Score perimeter 6 to 8 inches from the drip line.

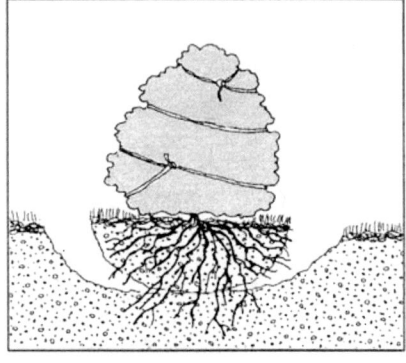

2 Tie up; cut, trim, shape root ball; undercut roots; tip root ball and slide moving material under it.

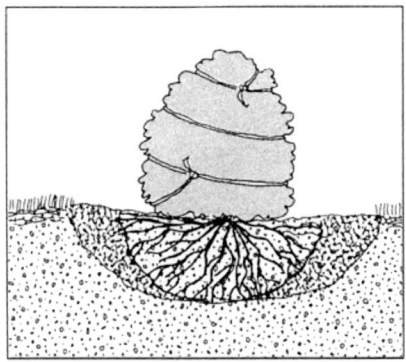

3 Set evenly in prepared hole; remove moving material; fill in with soil and gently but thoroughly soak in.

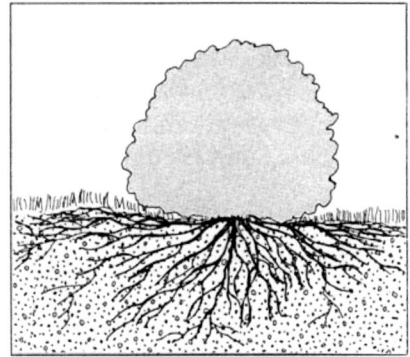

4 Typical anchor and feeder root system of a healthy boxwood plant.

Figure 5. Transplanting boxwood.

the drip line and score the perimeter with vertical cuts of the spade; tie up the plant; undercut the roots; cut, trim and shape the root ball; tip the root ball and slide burlap or other strong material under it; pin or tie the material around the root ball; do not lift the plant by its trunk or branches — drag it by the burlap or lift and carry with a suitable wheeled vehicle.

Dig the receiving hole as deep as the root ball and 6 to 8 inches wider. If the replanting is delayed for a few days, keep the transplant protected from the sun and wind; keep the roots and top moist.

Center the plant in the prepared hole and make certain the plant is no deeper than it was at its original site; withdraw the material covering the root ball.

Fill the remaining part of the hole with soil. Build a slight mound at the outer perimeter and untie the plant. Gently soak the plant in. Do not water every day — allow the soil to dry at the surface before watering again. Some judicious pruning of foliage is advised for the larger transplants. Remove up to $1/3$ of foliage if you suspect significant root damage or loss has occurred. You should expect yellowing of foliage during the first year after transplanting.

Container-Grown Plants

Container-grown boxwood are far easier to handle than the balled-and-burlapped specimens. Since a plant is sheathed in rigid plastic or metal, it is usually less vulnerable to damage while in transit. Plants can wait longer before planting and suffer less root disturbance in the course of planting/transplanting. Perfect as this may sound, container-grown boxwood should be given special treatment at planting time or they may never become well-established. A few general rules apply (Fig. 6):

1   Choose a healthy-looking plant; inspect the root ball; if medium is completely hidden by encircled roots, reject it.

2   Dig a hole at least twice as wide as the root ball and as deep; blend in an equal volume of organic material or compost.

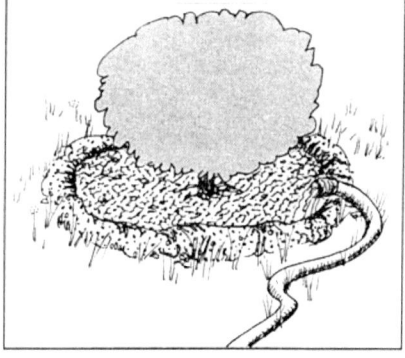

3   Prepare root ball by making four top to bottom cuts about 1 inch into the soil to sever any encircling roots; shake well.

4   Backfill around the root ball; tamp down soil so that no air spaces are left; build basin around to drip line; fill basin with water several times to thoroughly soak roots.

Figure 6.   Planting container-grown boxwood.

Cultivation

First, choose a healthy-looking boxwood. Check the leaves for any sign of insect infestation. Expose and inspect the root ball prior to purchase. Roots should be emerging from the planting medium. If the medium is completely obscured by a tangled mass of roots encircling the specimen, reject it.

Second, dig a hole at least twice as wide as the specimen's root ball and one and a half times as deep; blend in an equal volume of organic material or compost. Enrich the soil now or the roots may never expand beyond the original root ball. Container medium is usually a light, loose and organic material not well-suited to long-term root growth.

Third, prepare the root ball for planting by slipping or cutting it out of the container. Using a sharp knife, make four longitudinal cuts at even intervals around the root ball, slicing about an inch into the soil. This will sever any roots encircling the root ball that might gradually strangle it as it grows.

Fourth, before setting the root ball into the hole, shake it well to dislodge all loose nursery growing medium. Don't sink the roots deeper than they were in the container or they will ultimately be smothered. Backfill around the root ball, tamping down the soil so that no air spaces are left. Build a basin around the drip line to trap water to keep the roots moist. Fill the basin with water several times to thoroughly soak the roots. When watering, water slowly and thoroughly so that moisture penetrates all the way to the bottom of the root system.

Pruning and Renovation

Maintenance Pruning

Happily, pruning boxwood is very much a matter of personal philosophy. Most boxwood cultivars, when properly selected for landscape purposes, require little if any pruning except for the removal of dead wood and some very minor shaping. However, a few cultivars of dense and compact habit, as well as those prone to throwing sports (mutations), do require some regular pruning, thinning and sanitary cleanup as essentials of their cultural and aesthetic care.

Here are some general procedures and concepts that will cover most situations:

- Minor maintenance pruning is best done in the late spring after the new growth has started to harden off.
- Work from top to bottom of the plant and from the inside out. To set or control the plant's height, remove some top growth. To encourage compact growth, narrow the plant at the top to let sunlight and rainfall reach every part. When the plant grows unevenly, locate its center and prune for balance. This may require heavier pruning on one side than the other. If the new growth is rank and floppy, take it off and cut lightly into previous years' growth to encourage more compact new growth. Where branches are heavy, cut back severely, keeping in mind the natural form; repeat the following year.
- After the plant has been pruned, look inside and remove all broken branches, leaves and twigs. A good hard shake should dislodge most of the debris

caught in the branches, but any stubborn residue should be picked out. Do this job thoroughly one year and much of it can be skipped the next.

• If aesthetics are a requirement and major renovations are not necessary, don't remove more than one third of the branches in a year. If you must cut the top of a plant, also do some thinning in the body.

Major Renovation

The approach to major renovation is considerably different than routine pruning. It should be reserved for situations where boxwood has become seriously overgrown; impregnated with seedlings, vines and debris; and where edgings, hedgings and allées are so badly overgrown that "jungle" seems the best description.

The main tools are chain saws, arbor saws, heavy loppers, pruning shears and an eager crew.

Determine the desired size of edgings, hedgings and allées; then with the most advantageous tools, cut them back to at least 12 inches below the ultimate desired size. It may at first look like a horror of jumbled stubs, but within a few short years the plants will have fully recovered with only minor pruning maintenance required thereafter.

Large overgrown jungle areas can be advantageously rehabilitated by selecting the major trunks, cutting all others to the ground, then cutting off some lower branches of these major trunks to form a more tree-like shape.

Winter Damage — Protection and Corrective Action

Boxwood plants, like many other woody shrubs, are susceptible to winter damage. It is important to understand what may cause winter damage and to take necessary actions to prevent or at least minimize it.

Symptoms of winter damage:
- Foliage discoloration
- Defoliation
- Dead flower and leaf buds
- Broken branches, stems and twigs
- Cracks and splits along stems
- Sunken areas in trunk bark, crotches and main branches
- Death of entire branches and tops of the crown

Contributing factors:
- Selection of inappropriate cultivars for climatic conditions
- Inadequate soil moisture
- Fertilizing in late summer and early fall
- Excessively high and consistent winds during fall and winter
- High-low temperature fluctuations
- Out-of-season freezes

- Low temperatures that exceed normal hardiness range
- Heavy snow or ice loads
- Poor siting of foundation plantings
- Shoveling snow from walkways and driveways
- Chemical runoff

Protection:
- Select cultivars that fit your climatic area.
- Ensure that the boxwood enters the dormant season in a healthy and vigorous condition with adequate moisture.
- Fertilize boxwood in the late fall only (if you must).
- Provide wind protection for plants sited in areas exposed to strong or persistent winds. The use of snow fencing, lattice frames, or a loose covering with burlap are all good. These are particularly beneficial to recent transplants.
- Protect large boxwood against snow damage by wrapping the outer branches with strong soft cord. Tie the cord securely to a low branch, pressing the branches upward and inward while wrapping the cord about 8 to 10 inches apart. Have the cord just tight enough to help prevent breakage but not so tight as to preclude air circulation through the plant.
- Provide some additional mulch outside the drip line in late fall to help reduce the extremes in soil temperature fluctuations.
- It is better not to remove snow or ice from plants after a storm.
- Protect plantings in vulnerable snow slide areas by erecting temporary barriers over the plants. Such barriers will help break the fall of snow. Snow guards on the roof can also be helpful.
- Shovel snow from walkways and driveways any place but on your boxwood plantings.
- Exercise extreme caution in using salt and other chemical deicing agents.

Corrective action:
- If the boxwood has broken branches, dead stems and dead twigs, remove them by pruning back to live wood.
- If the foliage has turned yellowish or grayish-green, delay drastic pruning in the spring until new growth starts. Very often boxwood will produce new leaves in early spring; by late spring the injured leaves will have fallen and been replaced by new foliage. If not, pruning is in order. Some cultivars that are sited in full sun routinely turn a temporary reddish-brown during the winter.
- Proper cultural practices year round and the selection of appropriate cultivars still remain the best ways to prevent winter injury.

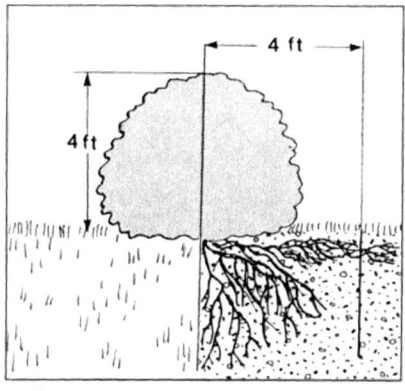

Typical growth of anchor and feeder roots in relation to plant size.

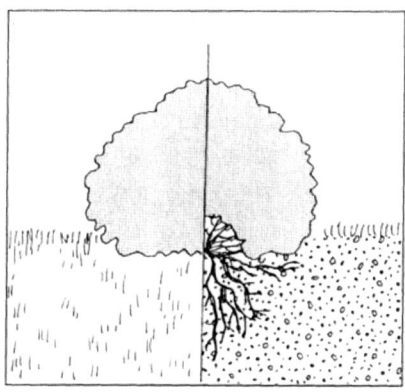

Leaf-debris accumulation and die-back of feeder roots.

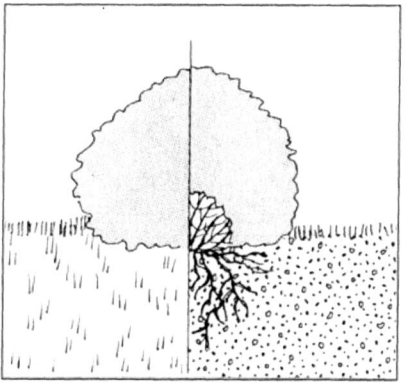

Leaf-debris accumulation with adventitious rooting and die-back of feeder roots.

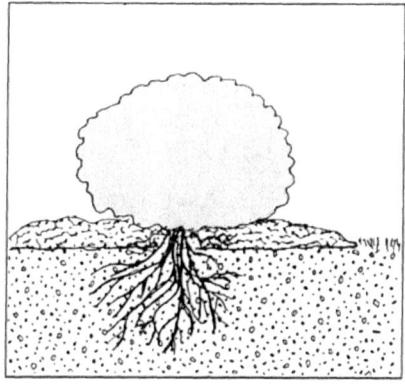

Excess mulch with adventitious rooting and die-back of feeder roots.

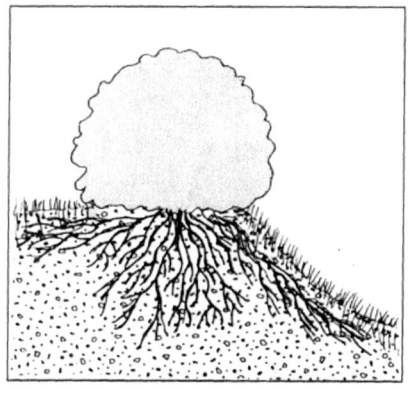

Feeder roots in well-stabilized sod on a slope.

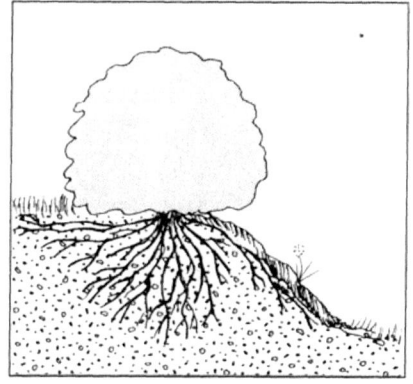

Feeder roots weakened by loss of grass and soil erosion.

Figure 7. Correct form and common problems in cultivated boxwood.

Cultivation

Antitranspirants and Antidesiccants

Many commercial suppliers advocate application of antitranspirants, also called antidesiccants, for the prevention of winter injury that may be caused by low temperatures and desiccation.

Based upon my own experience with boxwood and on my observations through the years, I am not convinced that antitranspirants are either beneficial or necessary for use on established plantings or transplants of boxwood.

Most Common Problems

Most boxwood problems are man-made. They make themselves known in the appearance of the whole plant, in the appearance of the leaves or roots, or in a combination. Most likely causes are summarized in Fig. 7:

Whole plant:
- Total lack of or excessive watering.
- Siting in water collection or drainage areas.
- Selection of wrong cultivar for the intended purpose or climatic conditions.
- Too much shade or sun.
- Salts and chemical runoff.
- Heavy concentrations of climbing vines.

Leaves:
- Boxwood leaf miners and mites.
- Crowding by other plants and structures.
- Inappropriate chemical drenchings and sprays.
- Dead leaves and debris accumulation in the sensitive cultivars.
- Sunscald on newly planted cultivars and cultivars sensitive to direct sunlight.

Roots:
- Planting too deep or settling later.
- Setting plants in heavy soils without providing drainage.
- Too much acidic material in the fill-in soil.
- Soil washing away from feeder roots.
- Digging or mechanical cultivation in feeder root areas.
- Application of excess amounts of synthetic fertilizers.
- Excessive amounts of mulch.

PROPAGATION

All boxwood species, varieties and cultivars do not respond in the same manner during the process of propagation. The following general information applies to some but not all. More specific details are provided for each species, variety and cultivar in the section covering Selected Species, Varieties, Cultivars.

Asexual Propagation (Cuttings)

Cuttings taken from mature boxwood may be rooted to produce young plants. At present, this is the method of choice to most likely produce plants identical to the parent plant. Unless you are using a natural dwarf or a small plant as a source, a boxwood plant can yield many cuttings without detriment to its health or form.

Cuttings may be taken in any season, but those taken between mid-June and mid-August usually root more successfully.

Select cuttings from the upper third of the plant, i.e., from "younger" plant tissue. Horticulturists indicate they are most apt to replicate the parent plant when taken from this position. Cuttings are best taken during the morning, not during mid-day heat or a period of drought.

From medium to large plants, select a healthy branch and make a cut of 6 to 8 inches from the tip of the selected branch. From dwarf or very small plants, remove 1 to 2 inches of healthy stem including a portion of heel from the previous year's growth.

Cuttings should be processed promptly. However, they may be held for several days wrapped in moist newspaper, cloth or sphagnum moss and then sealed in a plastic bag. Refrigeration (above freezing) helps hold specimens.

To encourage rooting, gently scratch or slit the surface of the stem in several places around the cut end; then gently remove all leaves from the lower third to half of the cutting. Avoid tearing the bark.

Rooting hormones are not essential but often hasten the rooting process for some cultivars. Commercially available products consisting of formulations of indole butyric acid, growth regulators, and fungicides such as Rootone® and Hormodin® have been used with success. Dip the prepared portion of the cutting in the hormone and tap gently to remove excess powder. The lower portion should be coated lightly and evenly.

The critical factor in rooting is high humidity. Alternative rooting environments for cuttings include a shady area outdoors under clear plastic enclosures or

Propagation

a greenhouse bed with a mist system. Pots enclosed in a plastic tent can be used quite successfully if a slight "misty" condition is maintained. Avoid heavy condensation in plastic tents because it encourages mold growth. Also avoid all exposure to sunlight; indirect lighting is most desirable.

Common rooting mediums are equal parts sand and peat, or peat and perlite, or commercial growing medium. To avoid disease buildup, use the rooting medium only once. Thoroughly premoisten and firm the medium before inserting a new cutting.

Use a pencil or stick to poke a hole in the rooting medium. Insert long (6 to 8 inches) cuttings to about 2 to 2½ inches in depth in the hole. Insert short cuttings up to about 1 inch in depth. Firm the medium around the stem, and water lightly to further settle the medium around the stems.

When taken during June through August, some cuttings root in about 6 to 8 weeks. More difficult cultivars may take longer (up to 5 or 6 months). Check the cuttings after 7 weeks by tugging on the stem very gently. When the root structure has formed, resistance to movement will be quite strong and the cuttings will be ready to pot. Transfer each rooted cutting to a 4- or 5-inch pot filled with coarsely textured potting soil.

Protect newly rooted plants during the first winter in a greenhouse, cold frame, cool room or similar environment. For the second winter, transfer to larger pots. Plunge the pots in the soil in a protected (especially from wind) area outdoors, water occasionally if necessary and mulch lightly.

Layering

Layering is accomplished by inducing a low-lying 1- or 2-year-old branch to produce an independent root system. The stem of the selected branch is first wounded by scraping its underside. Then bend down (do not cut) and bury the scraped branch in the soil to a depth of 2 or 3 inches. Place a brick on top of the buried section to keep the branch immobile; pegging or wiring into the ground may be necessary to achieve stability. In a year or two, when a root mass has formed on the previously scraped underside of the branch, sever the entire branch from the parent plant. Leave in place for several months before transferring the "new plant" to a lath house, shaded bed or pot. While layering will produce a plant identical to the parent, it does take a relatively long time.

Tissue Culture

Tissue culture is a relatively new laboratory technology of growing plants from individual plant cells in microculture; it may ultimately prove useful for boxwood. Much research remains to be done with this technique before marketable boxwood could be produced by this means.

Sexual Propagation

Seed

Plants grown from seed collected from mature plants will reflect some characteristics of each parent. Because the source of pollen is usually unknown, plants grown this way cannot be reliably associated with a given cultivar. Their hardiness, ultimate size and form cannot be predicted.

To collect open-pollinated seeds, cover an immature seed capsule with a cloth or paper bag. Do not use a plastic bag. At maturity the seed capsules will burst and release the seeds into the bag.

Plant boxwood seeds in a moist, well-drained, soil-less medium (available commercially) at about 65° F; hold in the dark until germination, which should occur in 6 to 8 weeks. Stratifying the seeds (i.e., storing seeds in a moist soil-less medium at about 40° F. for 10 weeks) may speed up and enhance germination.

When the seeds are placed in a cold frame or in a well-prepared protected seed bed they will usually germinate in the early to late spring, as the weather warms.

Once the true leaves are formed, transplant to a potting medium, such as $1/3$ peat, $1/3$ loam and $1/3$ perlite. Commercial mixes of growing medium are also satisfactory.

Controlled Hybridizing

This is a method of collecting pollen from male flowers on one plant and manually applying it to female flowers on another in order to produce seeds of known parentage.

It is necessary to bag the female flower before bud break with a cloth or paper bag (do not use a plastic bag) to prevent visits (and fertilization) by natural insect pollinators.

The two parent plants are usually selected for their desirable characteristics, and usually (but not always) their traits will be expressed in combination in the offspring.

This method is subject to trial and error and usually meets with mixed results. The traits of the offspring cannot be predicted and may not be expressed for many years. Given these problems, this has not been a useful method for generating new forms of boxwood.

Hybrid seeds are collected and germinated in the same manner as the seeds from open pollination, described above.

PESTS AND DISEASES

Most boxwood are free of pest problems. But if you do have a problem, this information should help in early detection and elimination.

The keys to pest control are knowing the enemy and being prepared to engage it. Each species of insect pest has a reasonably predictable pattern from birth to death. In the control process, timing is critical, because the emergence of pests varies with temperature, moisture, hours of daylight and local climatic conditions.

The boxwood leaf miner (*Monarthropalpus flavus*) and the boxwood mite (*Eurytetranychus buxi*) are currently the most serious pests of some boxwood species and cultivars (Johnson and Lyon, 1988). The boxwood psyllid (*Cacopsylla buxi*) is of lesser concern and causes no real injury unless heavy infestations are allowed to occur in successive years. Scientific names for pests follow those published by the Entomological Society of America (1989).

The Pests

The Boxwood Leaf Miner

Monarthropalpus flavus (Shrank), (Fig. 8) the boxwood leaf miner, is probably the most serious pest to some boxwood cultivars because it may defoliate the plant. Successive heavy annual infestations can even kill a plant. Some species, varieties and cultivars are more susceptible to this pest than others. None, however, are known to be immune (see the section Selected Species, Varieties, Cultivars).

The adult leaf miner is a small midge-like insect, orange to yellow-orange in color. Plant injury is recognized in late spring and early summer when small pinpoint blotchy areas, light gray in color, appear on the underside of new leaves. The blotches are visible to the naked eye, indicating the presence of eggs that have been deposited through the upper surface of the leaves. After hatching and feeding, the larvae overwinter *inside* the tissue of the leaf, and resume feeding in spring. Before pupation, the miners form small, paper-thin, gray caps on the underside of the leaves through which the adults emerge. Telltale pupal cases remain protruding through the exit holes after adults emerge.

In early spring as the temperature warms, boxwood begins to break dormancy, flower buds begin to open, and new leaf growth begins to show. This is the signal for the leaf miner to get into high gear. Larvae grow, pupate, adults emerge, fly, mate, lay eggs in the new leaf growth and die. In the very early spring, you should inspect the lower side of the leaves for evidence of last year's infestation. As the larvae begin to pupate, control measures should be taken at once: soil injection of Merit, or foliar spray with Merit or Avid, should give favorable results. This treatment can be effective in May, June and even July.

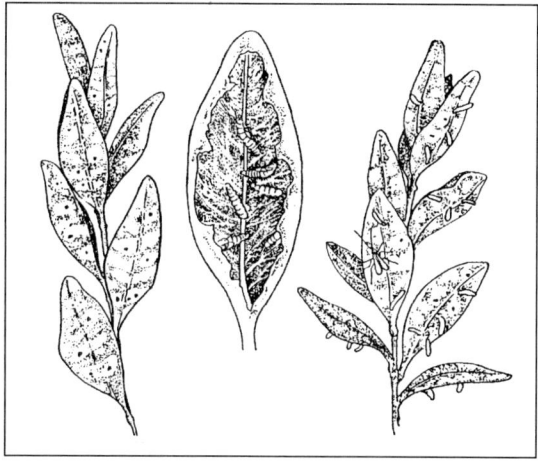

Figure 8. The Boxwood Leaf Miner

The Boxwood Mite

Eurytetranychus buxi, (Garman), (Fig. 9) the boxwood mite, is another serious pest, particularly in the more arid areas and during lengthy drought periods, when it tends to defoliate the plant. Some species, varieties and cultivars are more susceptible to this pest than others (see the section Selected Species, Varieties, Cultivars).

In appearance, the boxwood mite may vary from green to yellow-brown; the eggs are usually lemon yellow. To detect the presence of mites, shake several branches over a white cloth or paper. They will be easily visible as they crawl about. They feed by sucking the sap from leaves on both the upper and lower surfaces. When abundant, their feeding causes small white scratches on the leaves, giving them a blanched or silver-like appearance. This is followed by yellowing and premature leaf drop. During a hot dry summer, as many as seven generations of mites may occur.

To treat an infestation, first spray with a forceful stream of water to dislodge the mites from the leaf undersides. Repeat for three or four successive days. If mites are still present, use insecticidal soap every three to four days for two weeks. In the late fall, spray the previously treated boxwood with light horticultural oil to help destroy overwintering adults.

Figure 9. The Boxwood Mite

Pests and Diseases

The Boxwood Psyllid

Cacopsylla buxi (Linnaeus), (Fig. 10) the boxwood psyllid, is not usually a serious problem unless the plant is allowed to become heavily infested over several successive years.

The nymph overwinters in the egg stage under the scales of the boxwood flower bud. As the flower buds develop in the spring, the nymphs, which are visible to the naked eye, crawl from under the scales and develop into fly-like insects that lay their eggs in the new leaf growth. The feeding punctures of the greenish nymphs cause the new leaves to curl and cup. The deformed leaves are characteristic of psyllid infestation. Aside from being unsightly for the year, it may also reduce plant vigor.

In the very early spring, even in March, inspect the leaves for last year's infestation. If heavy, prune out the heaviest infestation and burn the affected branches. When the flower buds begin to swell and/or new leaf growth begins to show, spray with insecticidal soap every three or four days for two weeks. Intermittently brush the plant vigorously with your hands or a broom and this will dislodge many of the nymphs, which die before relocating the host plant. In the late fall, spray the previously treated boxwood with light horticultural oil to help destroy overwintering adults.

Figure 10. The Boxwood Psyllid

General Recommendations

Always apply sprays when weather conditions favor slow drying. Best times are morning, late afternoon, evening or during a period of heavy cloud cover. The longer the plant surface remains moist, the longer the spray will stay active in its liquid state. Avoid spraying on windy or gusty days. Also avoid periods of high temperature and humidity, which may increase the possibility of phytotoxicity.

Most insecticidal soap concentrates consist of potassium salts of fatty acids derived from plant and animal sources combined with water and alcohol. They are considered harmless to humans and are relatively benign in the environment. They must contact insects to be effective, as is the case with chemicals. Insecticidal soap is available at nurseries or through horticultural supply catalogs.

Any application of horticultural oil should not exceed a 2 percent solution in order for the oil to evaporate rapidly. A formulation for a slightly strong spray is 1 cup isopropyl alcohol and $1/2$ teaspoon horticultural oil mixed in 1 quart of water. Horticultural oil is available at nurseries or through horticultural supply catalogs.

Boxwood Diseases

It does little good to worry about diseases of boxwood. Research to isolate the causes and remedies have met with very little success. Effort is more profitably concentrated on proper cultural care, for prevention through health is the key. A plant in ill health usually suffers from a cultural problem — poor siting, poor drainage, etc. Correction and patience may bring recovery. There are no effective fungicides or nematicides to prevent or control the diseases that have been associated with boxwood. For those academically interested in boxwood diseases, the following information may prove helpful.

Boxwood pathogens include a complex of fungi such as forms of *Ganoderma, Volutella, Nectria, Phyllactinia* and *Pseudonectria* (Sinclair, Lyon and Johnson, 1987).

Reputed outbreaks include:

- *1900s – 1940s* Many boxwood enthusiasts were led to believe that their plants were doomed to canker from *Volutella* or fungal leaf and stem rot. After considerable debate and time, it was conceded that the diseases were manifestations of decline from improper cultural practices.

- *1940s – 1960s* The nematode scare gripped the attention of boxwood growers in North America. Some horticulturists still believe that many boxwood were killed by the very same nematode treatments that were recommended at the time.

- *1960s – present* "Boxwood decline" has been the general term applied to any boxwood problem thought to be connected with diseases incited by root and colonizing fungi occurring at the basal portion of the stem. The term has also been loosely applied to symptoms induced by parasites or environmental factors acting directly on twigs and foliage.

Volutella leaf and stem blight is a disease caused by *Volutella buxi*, a fungus that primarily attacks the stems. It results in defoliation and death of the infected stem. It usually is found on unhealthy plants but does occasionally attack a healthy one. Symptoms include cream to light-pink mealy growth on the leaf undersides. On the blighted stems, dome-like spore masses erupt through the epidermis. At varying distances below tips of infected branches, the stem may be girdled. Further investigation may reveal a dark brown or black canker.

Root rot may be caused by drought, over-fertilization and/or severe winter injury. Transplants may decline and die as a result of poor cultural practices or siting. Symptoms include poor growth and lackluster foliage, which ultimately yellows. Leaves on isolated stems or on the plant as a whole turn upward and leaf margins roll inward, suggesting drought-related symptoms. A dark discoloration of the wood at the base of the stem usually occurs for 2 or 3 inches above the soil line. When roots are examined, they are few and many are brown. Lack of a functioning root system is the result of fungal infection and precedes the foliage symptoms and death of the crown. There is no known chemical means available to homeowners to prevent the development and spread of root rot on boxwood. By the time the disease is diagnosed from visible symptoms, it is beyond control anyway. The best hope for non-commercial growers is to start with healthy plants and give them the best possible care. Provide water when necessary but avoid over-watering or excess fertilizing. Thin compact plants to allow better air circulation as fungus prefers cool, moist, dark areas.

Root rot can also be caused by nematodes within the root system. There are no known nematicides that can treat this form of root rot once it manifests itself.

For plants in decline or with blighted stems or foliage, pruning helps. Prune dead and infected stems back to healthy tissue. Disinfect pruning equipment frequently with household bleach (diluted 1:9 with water), or undiluted isopropyl alcohol. Do not dispose of diseased boxwood prunings in your compost heap, and do not leave diseased plant material exposed to wind. Burn the cuttings or bury them in a landfill or non-horticultural site.

If boxwood plants have died and disease is confirmed, don't replant boxwood in the same location. Improve the soil and carefully evaluate potential drainage problems.

NOMENCLATURE, DOCUMENTATION, REGISTRATION

Introduction

Two prominent research botanists, horticulturists and propagators at the U.S. National Arboretum Plant Science Research Division, T.R. Dudley and G.K. Eisenbeiss, wrote in 1971:

> Perhaps the greatest bone of contention for amateur and professional horticulturists and plantsmen is the seemingly never-ending changing of names of plants which are of horticultural and economic importance. Botanists and taxonomists often incur great wrath when they publish name changes and alterations, even though done in good faith and strictly within established rules and procedural boundaries. And even when names are changed with the strongest of justifications, it often takes years for the new names to be accepted and substituted in literature or conversation for the incorrect names. Difficulties accrued for decades between botanists and horticulturists before coming to terms with the development and adoption of feasible and commercially acceptable procedures for the names of cultivated plants.

A frequent source of confusion to the layman is the difficulty, possibly delusionary, of distinguishing between common, botanical, and cultivar names. A common name is a colloquial or local name applied to a plant native to a particular area. While common names may have some importance when applied to native plants of a limited area, the identical common name often applies to different species of plants which may occur in widely separated geographical areas. Conversely, the same species of plant may grow in different areas and have different common names. Since common names have little scientific basis in origin or usage, attempts to standardize nationally or internationally have been notably unsuccessful. In contrast, Latin botanical names were originated for the most part with great care, are universal in application, and are documented and published according to rules and recommendations of the prescribed International Botanical Code. Cultivar names are likewise distinct from and must not be used interchangeably with common names. It is imperative that common names be clearly identified as such to avoid fallacious application of common, botanical, and cultivar names.

Taxonomic Nomenclature

The binomial system of naming plants was instituted by the Swedish botanist/physician Carl von Linné, more commonly known by his Latinized name, Carolus Linnaeus. In 1753 he wrote *Species Plantarum,* which gave a full account of the then-known plant species with a binomial name for each. This document would become the bible of taxonomists, those who deal with identification, naming and classification of organisms.

Linnaeus based his systematic botany on giving each species of plant two Latin names, first the genus, e.g. *Buxus* (the generic name), followed by the specific epithet (species), e.g. *microphylla*. The genus was a Latin or Latinized noun; the species usually a Latin adjective often descriptive of the plant. Many specific names were bestowed in the 18th and 19th centuries; others were carried over from earlier times, thus the beginnings of some plant nomenclature confusions.

A genus has been described as a more or less closely related and definable group comprising one or more species that have more characteristics in common with each other than they do with other species from other genera within the same family.

A species may very well be the most important unit of classification, though some consider the term more of a concept than an absolute. L.H. Bailey defined a species as a kind of plant or animal, distinct from other kinds in marked or essential features, that has good characters of identification. Also it may be assumed to represent a continuing succession of individuals from generation to generation. He then went on to say that the term is incapable of exact definition, for nature is not laid out in formal lines. Actually, the term "species" refers to a concept, the product of each individual botanist's judgment.

The term "variety" used in the botanical sense constitutes a group or class of plants subordinate to a species. It is usually applied to individuals displaying rather marked differences in nature. The crux is that these differences are inheritable and should show in succeeding generations. Unfortunately, "variety" is often confused, and used interchangeably, with the term "cultivar" (a term coined by L.H. Bailey to represent "cultivated variety").

"Cultivar" is a relatively new term, one that has become increasingly more important in horticultural circles. M.A. Dirr (1990) defines it as a cultivated form of species clearly distinguished by characteristics (morphological, physiological, cytological or chemical) that, when reproduced (usually asexually), retain its distinguishing characteristics.

Thus, we now have a complete systematic method and procedure for describing a plant — family, genus, species, variety, cultivar — that can be reasonably understood and that depicts the same meaning to those within the plant-oriented disciplines.

Example
Buxaceae *Buxus microphylla japonica* 'National'

Family	Genus	Species	Variety	Cultivar
Latin name	Latin generic name	Latin specific name	Latin varietal name	Modern language name
|	|	|	|	|
Buxaceae	*Buxus*	*microphylla*	*japonica*	'National'

Typographic Rules

Along with the foregoing methods and procedures of plant descriptions, typographic conventions have also evolved.

The generic name is written with the first letter capitalized; in contemporary use the specific epithet or name is written in lower case even when it is derived from a proper noun. When these names are printed, they appear in italics indicating a foreign language, i.e. *Buxus microphylla*.

The style in which the name of a plant is written depends on whether the plant is considered to be a variety or cultivar. A variety, generally understood to be a plant growing in the wild, is entitled to a varietal name in Latin. It follows the generic and specific name and may be written as var. (abbreviation for variety) followed by its Latin varietal name in lower case italic, i.e., *Buxus microphylla* var. *japonica*. If the plant is a cultivar, it may be written in either of two ways: 1) cv. (abbreviation for cultivar) followed by the cultivar name in capitals and lower case, i.e. *Buxus microphylla* var. *japonica* cv. National, or 2) cv. may be eliminated and the name, in capitals and lower case, is enclosed with single quotation marks, i.e. *Buxus microphylla* var. *japonica* 'National.' The latter convention is followed here.

Registration

The process of recording and establishing correct and valid names for new cultivars of *Buxus* is governed by the International Code of Nomenclature of Cultivated Plants, 1982, with the American Boxwood Society being the appointed international registration authority for cultivated *Buxus*. The present registrar is Lynn R. Batdorf, curator at the U.S. National Arboretum, Washington, D.C. 20002.

The main objective of registration is to stabilize and standardize naming of new cultivars. This goal is accomplished by recording, documenting, and publishing the origin, the discoverer, namer and introducer of new cultivar names. The modest information still existing for older cultivar names, in many languages, is scattered in many nursery catalogs; also in horticultural and botanical literature. Such information is often of dubious authenticity, with the unfortunate result that the names of most individuals who discovered or originated boxwood selections are lost to posterity. Likewise, the history of introduction and origin of many cultivars and their names is so obscure that numerous nomenclatural conflicts may never be satisfactorily resolved. Registration gives permanent recognition to persons involved with the discovery and introduction of new cultivars by documentation and publication of their names. It also preserves the cultivar names and guarantees, in good faith, that the names of originators, selectors, or introducers will always be associated with their valid and registered cultivar names. The practice of registration prevents the confusion that results from duplication of cultivar names. It also prevents the application of two or more different names to a single cultivar. Registration clearly distinguishes cultivar names from botanical names and common names.

New cultivar names of cultivated *Buxus*, to be regarded as valid, must fulfill the requirements and precepts established by the International Code of Nomenclature of Cultivated Plants, 1982. Briefly summarized, they require that:

1. The name must consist of not more than three words.
2. The cultivar name may be in any modern language, but may not be in Latin or have a Latinized ending.
3. It must be published in a recognized publication, accompanied by a botanical description or reference to a previous description.

The code further strongly recommends that, when possible, an illustration should be provided with the description; that an illustration and/or herbarium specimen be deposited in an herbarium.

International registration of cultivar names is not intended to judge the ornamental or horticultural merits of the plant. Registrars are instructed to avoid any judgment regarding the plant other than those relating directly to the validity and the applicability of the cultivar name.

SELECTED SPECIES, VARIETIES, CULTIVARS

Selected Species and Varieties

B. austro-yunnanensis	50
B. bahamensis	51
B. balearica	52
B. bartletti	54
B. bodineri	55
B. colchica	56
B. hainanensis	57
B. harlandii	58
B. hebecarpa	60
B. henryi	61
B. himalayensis	62
B. ichangensis	63
B. latistyla	64
B. linearifolia	65
B. megistophylla	66
B. microphylla	67
B. microphylla variety japonica	68
B. mollicula	70
B. myrica	71
B. pubiramea	72
B. rugulosa	73
B. sempervirens	74
B. sinica	75
B. sinica variety insularis	76
B. stenophylla	78
B. wallichiana	79

Selected Cultivars

B. harlandii
   'Richard' 80

B. microphylla
   'Compacta' 81
   'Creepy' 83
   'Curly Locks' 84
   'Grace Hendrick Phillips' 85
   'Green Pillow' 86
   'Helen Whiting' 87
   'Henry Hohman' 88
   'Jim's True Spreader' 89
   'John Baldwin' 90
   'Kingsville' 91
   'Locket' 92
   'Miss Jones' 93
   'Quiet End' 94
   'Sport Compacta No. 1' 95
   'Sport Compacta No. 2' 96
   'Sunlight' 97
   'Sunnyside' 98
   'Winter Gem' 99
   Partial Descriptions 100

B. microphylla variety *japonica*
   'Green Beauty' 101
   'Morris Dwarf' 102
   'Morris Midget' 103
   'Nana Compacta' 104
   'National' 105
   Partial Descriptions 106

B. sempervirens
   'Abilene' 107
   'Agram' 108
   'Arborescens' 109
   'Argenteo-variegata' 111
   'Aristocrat' 112
   'Asheville' 113
   'Aurea Pendula' 114
   'Aureo-variegata' 115
   'Belleville' 117
   'Blauer Heinz' 118
   'Bullata' 119
   'Butterworth' 120
   'Clembrook' 121
   'Cliffside' 122

'Decussata' 123
'Dee Runk' 124
'Denmark' 125
'Edgar Anderson' 126
'Elegantissima' 127
'Fastigiata' 128
'Flora Place' 129
'Fortunei Rotundifolia' 130
'Glauca' 131
'Graham Blandy' 132
'Handsworthiensis' 133
'Handsworthii' 135
'Hardwickensis' 136
'Heinrich Bruns' 137
'Henry Shaw' 138
'Hermann von Schrenk' 139
'Holland' 140
'Hood' 141
'Inglis' 142
'Ipek' 143
'Joe Gable' 144
'Joy' 145
'Krossi-livonia' 146
'Latifolia' 147
'Latifolia Marginata' 148
'Liberty' 149
'Macrophylla' 150
'Maculata' 151
'Mary Gamble' 152
'Memorial' 153
'Myosotidifolia' 154
'Myrtifolia' 155
'Natchez' 156
'Newport Blue' 157
'Nish' 158
'Northern Find' 159
'Northland' 160
'Notata' 161
'Pendula' 162
'Ponteyi' 163
'Prostrata' 164
'Pullman' 165
'Pyramidalis' 166
'Pyramidalis Hardwickensis' 167
'Rochester' 168
'Rotundifolia' 169

'Salicifolia' 170
'Salicifolia Elata' 171
'Ste. Genevieve' 172
'Suffruticosa' 173
'Tennessee' 175
'Undulifolia' 176
'Vardar Valley' 177
'Welleri' 179
'West Ridgeway' 180
'Woodland' 181
'Yorktown' 182
'Zehtung' 183
Partial Descriptions 184

B. sinica variety *insularis*
    'Justin Brouwers' 190
    'Nana' 191
    'Pincushion' 192
    'Tall Boy' 193
    'Tide Hill' 194
    'Winter Beauty' 195
    'Wintergreen' 196
    Partial Descriptions 197

B. ×
    'Green Gem' 198
    'Green Mound' 199
    'Green Mountain' 200
    'Green Velvet' 201

Key to Plant Descriptions

The plants in this section are arranged alphabetically by species and within species, with one plant type (taxon) discussed per page.

Plant: Only botanical names are used.

Size (25 yrs): A difficult subject and almost any designation can be challenged because of variations of climate, soils and siting. Every effort has been made to cite the size of actual living plants.

Criteria:
Dwarf: up to 2½ feet
Small: 2½ to 3½ feet
Medium: 3½ to 6 feet
Large: larger than 6 feet

Natural Form: The plants have been described regarding their general growth habit when only minor clipping and pruning have been performed to maintain the health of the plant. Line drawings of the natural forms at approximately 25 years of age are provided for many of the plants.

Criteria:
Arboreal: like a tree
Columnar: like a column or telephone pole
Conical: cone-shaped
Mounded: semi-globular, wider than tall
Ovate: egg-shaped outline
Pyramidal: broadest at base, tapering to apex
Spherical: rounded body
Unusual: forms not fitting the above descriptions
Vase: tapering from small base to larger top, generally single-trunked

Annual Growth Rate: The growth rate data are also influenced by the variables of climate, soils, water, fertility and siting. The annual growth rates are those that can be expected under "normal" situations.

Criteria:
Slow: up to 1½ inches
Medium: 1½ to 3½ inches
Fast: more than 3½ inches

Leaf Color: Quite subjective. Seasons, soils, lighting and siting all affect color to varying degrees.

Criteria:
Light, medium or dark green
Light yellow-green, medium yellow-green or dark yellow-green

Selected Species, Varieties, Cultivars

Leaf Shape:
Criteria:
Leaf shape:
Elliptic: oval, widest at the middle
Lanceolate: lance-shaped
Oblanceolate: opposite of lance-shaped
Obovate: opposite of lance-shaped, widest above the middle
Ovate: egg-shaped, widest below the middle
Rotund: rounded
Revolute: rolled backward at the tip or margin
Leaf tip:
Acute: narrowed to a point
Obtuse: rounded at the apex
Retuse: notched at the tip

Leaf Size: Life-size twig-leaf illustrations are included, as well as measurements of both leaf length and width.
Criteria:
Small: up to $3/4$ inch long
Medium: $3/4$ to $13/16$ inch long
Large: more than $13/16$ inch long

Leaf Surface: Quite subjective, but describes surfaces in terms of those most common to boxwood.
Criteria:
Bullate: puckered or blistered
Glabrous: smooth and not hairy
Glaucous: with a powdery coating
Glossy: shiny, smooth, not hairy
Matte: dull, flat

Internodal Length: The distance between the leaf nodes.
Criteria:
Short: up to $1/4$ inch
Medium: $1/4$ to $13/16$ inch
Long: more than $13/16$ inch

Flowering Habit: Most boxwood that flower are not conspicuous enough to be of significant ornamental value. However, many boxwood do flower and thus add another element to plant description.
Criteria:
Floriferous: heavy and conspicuous
Moderate: medium and somewhat conspicuous
Sparse: little flowering and not very obvious
Not observed: have never seen and/or available literature does not indicate

Hardiness: Hardiness zone designations are not absolute. Many factors other than temperatures affect plant survival in a specific area. While several hardiness zonal charts have been published, I have selected the latest available U.S. Department of Agriculture plant hardiness map (Fig. 11) for use herein. Hardiness rating is meant only as a guide and is not necessarily a limiting factor in plant use. Generally, boxwood hardiness range is between Zone 5 (-20°) and Zone 10 (40°). Within reason, I considered every boxwood to be hardy unless I've killed it myself.

Criteria:
Approximate range of average annual minimum temperatures

Zone	Fahrenheit
5	-20° to -10°
6	-10° to 0°
7	0° to 10°
8	10° to 20°
9	20° to 30°
10	30° to 40°

Nearly all of North America that lies between USDA plant hardiness Zones 5 through 9 can readily be called boxwood country, providing that the appropriate cultivars are selected to coincide with the local climatic conditions.

Some would proclaim boxwood country to correspond only to USDA hardiness Zone 6. This is true in the sense that the largest number of cultivars grow in this zone. However, a good number of these species, varieties and cultivars thrive in Zones 5 through 9 as well, when given proper care.

Landscape Use: Purely a judgment as to where in the landscape a plant can be effective. Criteria reflect those usages known to the author, and are in no way intended to be exclusive.

Criteria:

Allée	Grouping for background
Area separations	Hedging
Bonsai	Knot garden
Container plantings	Maze
Edgings	Parterres
Espalier	Specimen
Foundation plantings	Topiary

Plant Registration: This indicates when proper evidence for international registration of a cultivar was presented to the International Registration Authority for Cultivated *Buxus* L., and the official document in which it was published. It also identifies the person or agency that provided the evidence, and establishes when the action was taken. Many cultivars have not been registered; nonetheless, they are included as having importance and in due course may qualify for registration.

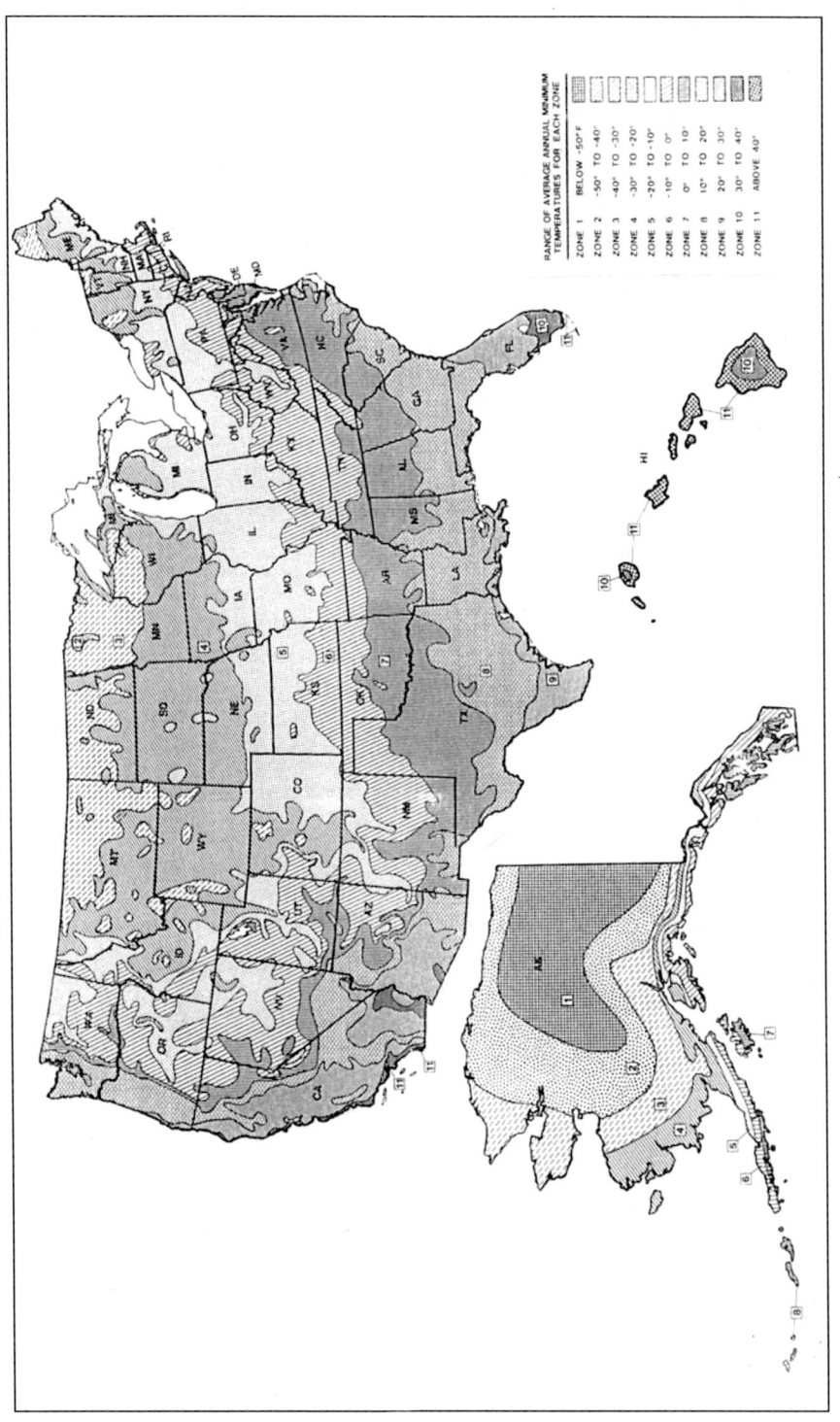

Figure 11. USDA plant hardiness zone map.

History: There are many interesting stories about plant discoverers, propagators, introducers and namers of boxwood that lend credence and authenticity to the plant character and origins.

Additional Information: When the selected plants do not fit the basic cultivation information provided in the following text, the exceptions will be noted.

Culture and Care: Most boxwood cultivars transplant readily, and generally prefer dappled shade; most will tolerate but not always enjoy being sited in direct sun. Adding compost to the soil and one inch of mulch improves the health of the plant. Water seldom and thoroughly to augment your particular rainfall pattern. An inch of water every week is sufficient for an established plant in well-drained soil. Most boxwood readily tolerate a pH range of slightly acidic (6.5) to slightly alkaline (7.2) with a preference for the sweet side (alkaline).

Pests and Diseases: Only the leaf miner, spider mite and psyllid are currently considered potentially serious threats to some cultivars. Some boxwood cultivars exhibit natural resistance to these pests and will be so identified. All the current pests are controllable. Diseases are practically non-existent and are usually from man-made causes.

Propagation: All boxwood do not respond in the same manner during the process of propagation. Most propagate vegetatively quite readily and those requiring special attention will be noted.

Nearly all the selected plants are available from the commercial nursery trade. The American Boxwood Society publishes *Boxwood Buyer's Guide* listing many of the commercial nurseries and their available boxwood species, varieties and cultivars.

Buxus austro-yunnanensis – Species

Size (25 yrs.):	Large – 6 to 8¼ feet in height.
Natural Form:	Unusual – prostrate to upright.
Annual Growth:	Fast – 3½ to 4 inches in height.
Leaf Color:	Not observed.
Leaf Shape:	Oblanceolate to narrowly obovate; obtuse tip and slightly retuse; cuneate base.
Leaf Size:	Large – ⁷/₁₆ to 1 inch long and ³/₁₆ to ⁵/₁₆ inch wide.
Leaf Surface:	Glabrous and smooth with veins quite distinct.
Internodal Length:	Medium – ¼ to ⅜ inch.
Flowering Habit:	Medium flowering and fruiting.
Hardiness:	Not observed.
Plant Use:	Specimen.

Registration: Hatusima in *Journ. Dept. Agr. Kyushu Univ.*, 6(6):286, f.7, a-d, Pl. 17(2), f.1, 1942; *Yunnan Journal of Plants* 1:142, Pl. 35, 4-7, 1977; Zheng Mian, Min Tianly in *Flora Reipublicae Popularis Sinicae, Tomus* 45 (1), Science Press, 1980.
History: Found in southern Yunnan (Shangjiang, Lancang, Jinghong); grows beside rivers in fissures in rocks or in thick bush-wood, altitude 480-890 m. Type specimen collected near Jinghong, Yunnan, China.
Bibliography:
The Boxwood Bulletin, Vol. 30(4):67, April 1991.
Known Locations: Not documented
Additional Information: Not documented

Buxus bahamensis – Species

Size (25 yrs.):	Large – 6 to 7 feet high and 5 to 6 feet wide.
Natural Form:	Arboreal.
Annual Growth:	Medium – $2^1/_2$ to 3 inches in height and 2 to $2^1/_2$ inches in width.
Leaf Color:	Medium yellow-green.
Leaf Shape:	Elliptic, with some being slightly twisted; acute tip; cuneate base.
Leaf Size:	Large – 1 to $1^1/_4$ inches long and $^3/_8$ to $^5/_{16}$ inch wide.
Leaf Surface:	Glabrous and smooth.
Internodal Length:	Long – $^3/_8$ to $^5/_{16}$ inch.
Flowering Habit:	Moderate flowering and moderate fruiting.
Hardiness:	Zones 9 to 10.
Plant Use:	Specimen.

Registration: Baker in Hooker's *Icones Plantarum*, Plate 1806. 1889.

History: *B. bahamensis* is found on many islands of the Bahama Islands, Cuba and Jamaica, growing only in dry rocky locations where it is exposed to occasional salt spray, a high pH and a hot sun. Dr. John Popenoe, director of the Fairchild Tropical Garden in Miami, Fla., says:

"Bahama boxwood, *Buxus bahamensis*, is a tropical relative of our common ornamental boxwoods. It is an evergreen shrub or small tree native to the Bahama Islands, Cuba and Jamaica. It is not a plant of tropical rainforests, however, and grows only in dry rocky locations where it is exposed to occasional salt spray, a high soil pH and a hot sun."

He further writes in the *American Boxwood Society Bulletin* that:

"There is only one species of *Buxus* in the Bahamas, but the Cuban flora lists 27 for that island. Bahama boxwood, if we may use this as a common name, is found on many islands of the archipelago, where I have seen it from New Providence to Inagua. On Inagua, this plant has the common name of parrot wood; I have heard no common name for it on other islands. It is commonly crowded in among scrub growth just back from the seashore and does not distinguish itself in any particular way. The foliage is generally compact and green to yellow-green."

Bibliography:
Baker in Hooker's *Icones Plantarum*, Plate 1806. 1889.
The Boxwood Bulletin, Vol. 9(2):22-23, Oct. 1969.

Known Locations: Not documented

Buxus balearica – Species

Size (25 yrs.):	Large – 7 to 8 feet high and 4 to 5 feet wide.
Natural Form:	Arboreal and tending toward conical.
Annual Growth:	Medium – 2 to 2½ inches in height and 1½ to 2 inches in width.
Leaf Color:	Dark green.
Leaf Shape:	Elliptical to revolute; broadly acute tip with some being retuse; cuneate base.
Leaf Size:	Large – 1 to 1⅞ inches long and ½ to ¾ inch wide.
Leaf Surface:	Glabrous – tending toward glossy and quite leathery.
Internodal Length:	Long – ⅜ to 7/16 inch.
Flowering Habit:	Floriferous and heavy fruiting; female flowers sessile and male flowers stalked.
Hardiness:	Zones 7(protected) to 10.
Plant Use:	Specimen, grouping for background and area separations.

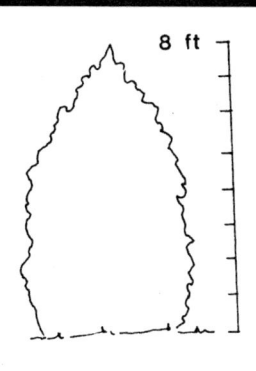

Registration: Willdenow in *Species Plantarum*, 4:337 - Lamarck, *Encyc. Meth. Bot.*, 1.511.1785. Synonym: *Buxus sempervirens* variety *gigantea*.
History: Native to the Balearic Islands of Minorca and Majorca, Sardinia, Corsica, Asia Minor and Turkey. It is often referred to as Turkey Box or Minorca Box. It came to France about 1770 and on to England about 1780. North America received its first *B. balearica* in the early 1900s. It was mentioned twice in the Bible (Isaiah 30:8 and 41:19). Theophrastus, who lived in the 3rd century B.C. and who is known as the father of botany, mentioned it, as had the Grecian poet Homer some five or more centuries earlier. There are references to the use of the fine hard wood in shipbuilding, in the making of small boxes, of mathematical and musical instruments. *B. balearica* is reported by Cimonent and Miedzyrzecki (1932) to have a haploid number of $n=14$.
Bibliography:
Bailey, L.H. *Hortus Third*, 1976.
Bean, W.T. *Trees and Shrubs Hardy in the British Isles.*
Beckett, K.A. *The Complete Book of Evergreens*, Van Nostrand Reinhold Co., 1981.
Dallimore, W. *Holly, Yew and Box*, 1908.
Everett, T.H. *The New York Botanical Garden Ill. Encyc. of Hort.*, Vol. 2, 1981.
Krussman, G. *Manual of Cultivated Broad-leaved Shrubs*, Vol., A-D, 1984.
Loudon. *Trees and Shrubs*, 1875.

The Boxwood Bulletin, Vol. 1(2):Cover, ifc, Jan. 1962/2(3):37-38, Jan. 1963/3(2):25, Oct. 1968/8(2):25-26, Oct. 1968/14(1):13, July 1974/16(1):2, July 1976/17(2):25, Oct. 1977/23(4):25, April 1984/26(1):12-15, July 1986.

Known Locations: College of William and Mary, Missouri Botanical Garden, Washington Park Arboretum, U.S. National Arboretum, State Arboretum of Virginia.

Additional Information:
Culture and Care: Transplants readily; will tolerate being sited in some direct sun but occasionally suffers from winter bronzing. The addition of organic compost as a soil amendment and one inch of mulch adds to the health of the plant. Water seldom and thoroughly; an inch of water every week is sufficient for sites with well-drained soil. Tolerates a pH range of slightly acidic to slightly alkaline with a preference for the sweet side (alkaline). Demonstrates no special cultural requirements.

Pests and Diseases: Indicates resistance to leaf miner, psyllid and mites in the more humid climates; no serious diseases.

Propagation: Cuttings root quite readily without the use of an IBA powder dip; however, they root slightly faster with the dip. The poly-tent procedure usually produces rooted cuttings nearly as fast as the mist systems, about 6 to 8 weeks.

Available in the commercial nursery trade in North America and Europe.

Buxus bartletti – Species

Size (25 yrs.):	Large, occasionally to 15 feet.
Natural Form:	Arboreal, quite loose and open.
Annual Growth:	Fast.
Leaf Color:	Medium green.
Leaf Shape:	Elliptic; acute tip; cuneate base.
Leaf Size:	Large – $3/4$ to $1 1/4$ inches long and $1/4$ to $1/2$ inch wide.
Leaf Surface:	Glabrous and smooth.
Internodal Length:	Medium – $3/8$ to $1/2$ inch.
Flowering Habit:	Not observed.
Hardiness:	Not observed, most likely Zone 10.
Plant Use:	Specimen.

General Comment: Illustration not available.
Registration: Standley in Publication 350, Field Museum of Natural History. Chicago, Botanical Series 11:134, 1932.
History: In February 1931, the late Professor H.H. Bartlett of the University of Michigan collected a plant along the river bluffs of El Cayo District, British Honduras, that the late Paul C. Standley of the Field Museum of Natural History (now Chicago Natural History Museum) described under the name *Buxus bartletti*. This was apparently the first representative to be collected in Central America. Later collections have been made in Mexico. Samuel J. Record and Robert W. Hess in *Timbers of the New World* (Yale University Press, 1943) state as follows:

"Tricera, either as a distinct genus or as a section of *Buxus*, includes about 15 species having their center of distribution in the West Indies, with known extensions into southern Mexico, British Honduras, Panama and Venezuela. They are rather rare plants, nearly all evergreen shrubs, though occasionally trees 15 to 25 feet tall. The wood is light yellow throughout; odorless and tasteless; hard, heavy, compact, of very fine and uniform texture; closely resembles *Buxus sempervirens*. It is not of commercial value because of the scarcity of the larger sizes, but it is suitable for articles of turnery and for engraving."

Bibliography:
The Boxwood Bulletin, Vol. 9(2):23-24, Oct. 1969.
Known Locations: Not documented

Buxus bodineri – Species

Size (25 yrs.):	Medium – 4 to 4½ feet high and 4 to 5 feet wide.
Natural Form:	Somewhat pyramidal.
Annual Growth:	Medium – 2 to 2¼ inches in height and 2 to 2½ inches in width.
Leaf Color:	Medium green.
Leaf Shape:	Oblanceolate; obtuse tip with some being slightly retuse; cuneate base.
Leaf Size:	Large – 1¼ to 1½ inches long and 5/16 to 3/8 inch wide.
Leaf Surface:	Glossy and smooth.
Internodal Length:	Medium – ½ to 9/16 inch.
Flowering Habit:	Moderate.
Hardiness:	Zones 6 to 8.
Plant Use:	Specimen.

Registration: Léveillé in, Feddes *Repertorium* 11:549, 1913; Kew S.r., 44:450R27.
History: Originated in China, Kweichau province.
Bibliography:
Everett, T.H. *The New York Botanical Garden Ill. Encyc. of Hort.*, Vol. 2, 1981.
Mian, Z. and Tianly, M. *Flora Reipublicae Popularis Sinicae, Tomus* 45(1), 1980.
The Boxwood Bulletin, Vol. 26(1):20, July 1986/31(1):17-18, July 1991.
Known Locations: Royal Botanic Gardens, Edinburgh; Royal Botanic Gardens, Kew, Richmond, England.

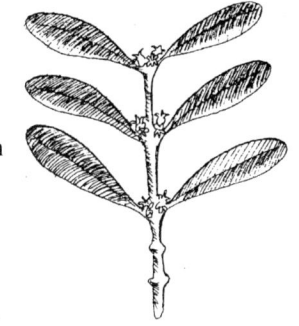

Buxus colchica – Species

Size (25 yrs.):	Medium – 3½ to 4 feet in height and 4 to 5 feet in width.
Natural Form:	Somewhat pyramidal.
Annual Growth:	Medium – 1½ to 2 inches in height and 2 to 2¼ inches in width.
Leaf Color:	Medium green.
Leaf Shape:	Elliptic; acute tip; cuneate base.
Leaf Size:	Medium – ¾ to ⅞ inch long and ⅜ to 5/16 inch wide.
Leaf Surface:	Glabrous and smooth.
Internodal Length:	Medium – ⅜ to 5/16 inch.
Flowering Habit:	Not observed.
Hardiness:	Zones 6 to 8.
Plant Use:	Specimen.

Registration: Pojark, in *Referat. Nauch-Issl. Rab. Akad. Nauk. SSSR*, Biol. 1945, 7, in obs.
History: Native to the Caucasus region.
Bibliography:
The Boxwood Bulletin, Vol. 16(1):12-13, July 1976/27(3):65, Jan. 1988.
Known Locations: Washington Park Arboretum, U.S. National Arboretum, State Arboretum of Virginia.

Buxus hainanensis – Species

Size (25 yrs.):	Large – 5 to 7 feet in height.
Natural Form:	Not observed – described as a shrub.
Annual Growth:	Medium.
Leaf Color:	Not observed.
Leaf Shape:	Elliptical – oblong or oblong-lanceolate; leaf tip attenuate, obtuse or with a small point slightly protruding; cuneate base.
Leaf Size:	Large.
Leaf Surface:	Glabrous and smooth.
Internodal Length:	Long – more than $^{13}/_{16}$ inch.
Flowering Habit:	Medium flowering and fruiting.
Hardiness:	Not observed.
Plant Use:	Specimen.

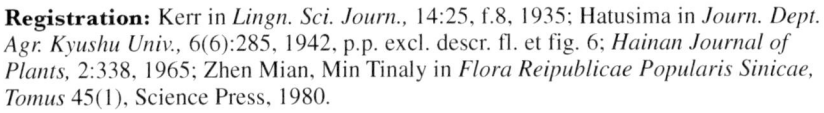

Registration: Kerr in *Lingn. Sci. Journ.*, 14:25, f.8, 1935; Hatusima in *Journ. Dept. Agr. Kyushu Univ.*, 6(6):285, 1942, p.p. excl. descr. fl. et fig. 6; *Hainan Journal of Plants*, 2:338, 1965; Zhen Mian, Min Tinaly in *Flora Reipublicae Popularis Sinicae, Tomus* 45(1), Science Press, 1980.
History: Found in Guandong (Ya County in Hainan Island); grows by streams or under moist wooded cover. Type specimen collected from Huangjin Mountain, Ya County, Hainan Island, China.
Bibliography:
The Boxwood Bulletin, Vol. 30(3):53, Jan. 1991.
Known Locations: Not documented
Additional Information: Not documented

Buxus harlandii – Species

Size (25 yrs.):	Medium – 4½ to 5 feet high and 5 to 6 feet wide.
Natural Form:	Vase-shaped; another form more spherical and slightly pendulous.
Annual Growth:	Medium – 2 to 2¼ inches high; 2¾ to 3 inches wide.
Leaf Color:	Medium green.
Leaf Shape:	Narrowly obovate to oblanceolate; obtuse tip with some being slightly retuse and curled inward; cuneate base.
Leaf Size:	Large – 1½ to 1¾ inches long and 5/16 to 3/8 inch wide.
Leaf Surface:	Glossy – smooth.
Internodal Length:	Medium – 5/16 to 3/8 inch.
Flowering Habit:	Sparse flowering and fruiting.
Hardiness:	Zones 6 to 8.
Plant Use:	Specimen, foundation planting, grouping for background and area separations.

Registration: H. Hance, Supplement to the Flora Hongkongensis in *Journal of Linnean Society*, 13:123-124, 1873.

History: The plant from which specimen plants found in America were derived was apparently collected by E.H. Wilson in west Hupeh, China: "Ichang gorge, on rocks, alt. 30-300 m. March 24, 1908 (No. 3399); 15-30 cm tall. Herbarium Arnold Arboretum." Dr. J.T. Baldwin, Jr. brought to the College of William and Mary a second clone from Hong Kong having the same growth characteristics but more graceful in habit. He also brought another clone from Royal Botanic Gardens, Kew, in 1952 that was labeled *Buxus nepalense,* which grows into a vase shape. Some believe it to be an invalid classification.

Bibliography:
Bailey, L.H. *Hortus Third*, 1976.
Dallimore, W. *Holly, Yew and Box*, 1908.
Dirr, M.A. *Manual of Woody Landscape Plants*, 4th Ed., 1990.
Everett, T.H. *The New York Botanical Garden Ill. Encyc. of Hort.*, Vol. 1, 1981.
Krussman, G. *Manual of Cultivated Broad-leaved Trees and Shrubs*, Vol. I, A-D, 1984.
Mian, Z. and Tianly, M. *Flora Reipublicae Popularis Sinicae, Tomus* 45(1), 1980.
Taylor, N. *Taylor's Encyclopedia of Gardening*, 3rd Ed.

Wyman, D. *Wyman's Gardening Encyclopedia,* New Expanded 2nd Ed., 1986.
The Boxwood Bulletin, Vol. 2(3):38, Jan. 1963/2(4):44, April 1963/6(3):40-42, Jan. 1967/12(1):16, July 1972/13(3):45, Jan. 1974/14(1):13, July 1974/17(2):25, Oct. 1977/22(4):71, April 1983/27(3):65, Jan. 1988/28(1):3-8, July 1988.

Known Locations: Brooklyn Botanic Garden, College of William and Mary, U.S. National Arboretum, Washington Park Arboretum, State Arboretum of Virginia.

Additional Information:
Culture and Care: Prefers dappled shade and to be out of the winter winds. Breaks dormancy early, prone to dieback from late freezes. Dieback not injurious, just ugly. Prune to clean up the appearance.
Pests and Diseases: Resistance to leaf miner, psyllid and mites.
Propagation: Bruise the stem tissue, use an IBA powder dip.
Available in the commercial nursery trade.

Buxus hebecarpa – Species

Size (25 yrs.):	Large – more than 6 feet in height.
Natural Form:	Described as a shrub.
Annual Growth:	Medium – in excess of 2 inches per year.
Leaf Color:	Not observed.
Leaf Shape:	Elliptic or oblong-ovate, rarely lanceolate; acute tip; cuneate base.
Leaf Size:	Large – 1 to 1½ inches in length and ½ to ⅝ inch in width.
Leaf Surface:	Glabrous and smooth, midrib very prominent.
Internodal Length:	Medium – ⅝ to 1 inch.
Flowering Habit:	Medium flowering and fruiting.
Hardiness:	Not observed.
Plant Use:	Specimen.

Registration: Hatusima in *Journ. Dept. Agr. Kyushu Univ.*, 6(6):302, f.14, 1942; Zheng Mian, Min Tianly in *Flora Reipublicae Popularis Sinicae, Tomus* 45(1), Science Press, 1980.
History: Found in Mount Emei in Sichuan and Tianquan Erlang mountains; grows in forests or rocks, altitude 1600-2000 m. Type specimen collected in Mount Emei.
Bibliography:
The Boxwood Bulletin, Vol. 30(4):69-70, April 1991.
Known Locations: Not documented
Additional Information: Not documented

Buxus henryi – Species

Size (25 yrs.):	Not observed.
Natural Form:	Not observed.
Annual Growth:	Not observed.
Leaf Color:	Not observed.
Leaf Shape:	Narrowly elliptic to narrowly lanceolate; acute tip and somewhat revolute; cuneate base.
Leaf Size:	Medium – 3/4 to 1 inch long and 3/16 to 1/4 inch wide.
Leaf Surface:	Glabrous and smooth.
Internodal Length:	Medium – 1/4 to 3/8 inch.
Flowering Habit:	Moderate flowering and moderate fruiting; flowers stalked with erect style and reflexed stigma.
Hardiness:	Zones 6 to 8.
Plant Use:	Specimen.

Registration: H. Mayr in *Fremdlandisch Waldund Sparkbaume fur Europa* 451, 1906.
History: Discovered in west Hupeh, China.
Bibliography:
Krussman, G. *Manual of Cultivated Broad-leaved Trees and Shrubs*, Vol. I, A-D, 1986.
Known Locations: Not documented

Buxus himalayensis – Species

Size (25 yrs.):	Large – 8 to 9 feet high and 6 to 7 feet wide. A 30-year-old plant is 10½ feet high and 9 feet wide.
Natural Form:	Pyramidal, billowy and compact.
Annual Growth:	Fast – 3½ to 4 inches in height and 2½ to 3 inches in width.
Leaf Color:	Medium green.
Leaf Shape:	Elliptic; acute tip; cuneate base.
Leaf Size:	Medium – ¾ to 1⅛ inches long and ⁵/₁₆ to ⅜ inch wide.
Leaf Surface:	Glabrous and smooth.
Internodal Length:	Medium – ½ to ⅝ inch.
Flowering Habit:	Sparse flowering and sparse fruiting.
Hardiness:	Zones 6 to 8.
Plant Use:	Specimen, grouping for background and area separations, allees, topiary.

Registration: Not registered.
History: Discovered in the northwest region of the Indian Himalayas.
Bibliography:
Krussman, G. *Manual of Cultivated Broad-leaved Trees and Shrubs*, Vol. I, A-D, 1984.
The Boxwood Bulletin, Vol. 16(1): July 1976/26(2):26-27, Oct. 1986/27(4):80, April 1988.
Known Locations: U.S. National Arboretum, State Arboretum of Virginia.

Buxus ichangensis – Species

Size (25 yrs.):	Medium – 3½ to 4 feet in height.
Natural Form:	Not observed – described as a shrub.
Annual Growth:	Medium – 1½ to 2 inches in height.
Leaf Color:	Not observed.
Leaf Shape:	Oblanceolate; obtuse tip; cuneate base.
Leaf Size:	Medium – ¾ to 1 inch long and 3/16 to ¼ inch wide.
Leaf Surface:	Glabrous and smooth.
Internodal Length:	Medium – ¼ to 5/16 inch.
Flowering Habit:	Medium flowering and fruiting.
Hardiness:	Not observed.
Plant Use:	Specimen.

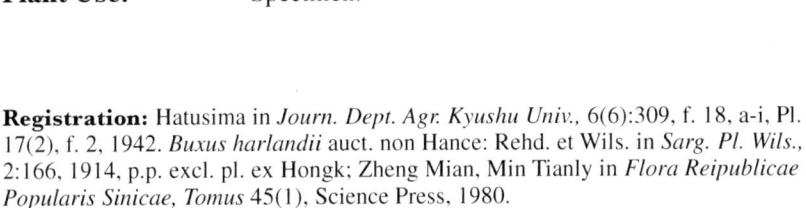

Registration: Hatusima in *Journ. Dept. Agr. Kyushu Univ.*, 6(6):309, f. 18, a-i, Pl. 17(2), f. 2, 1942. *Buxus harlandii* auct. non Hance: Rehd. et Wils. in *Sarg. Pl. Wils.*, 2:166, 1914, p.p. excl. pl. ex Hongk; Zheng Mian, Min Tianly in *Flora Reipublicae Popularis Sinicae, Tomus* 45(1), Science Press, 1980.
History: Found in western Hubei (the region of Badong, Zigui and Yichang); grows on river banks or on rocks facing the sun, altitude 30-300 m. Type specimen collected in Badong, Hubei.
Bibliography:
The Boxwood Bulletin, Vol. 31(1):19, July 1991.
Known Locations: Not documented
Additional Information: Not documented

Buxus latistyla – Species

Size (25 yrs.):	Large – 3½ to 13 feet in height.
Natural Form:	Described as a shrub.
Annual Growth:	Fast.
Leaf Color:	Not observed.
Leaf Shape:	Ovate or oblong-ovate; acute tip, obtuse or with a small point protruding; base round or very obtuse, margin curves downward, midrib prominent on both sides of leaf.
Leaf Size:	Large – 1 to 2 inches in length and ½ to 1 inch wide.
Leaf Surface:	Glabrous and smooth.
Internodal Length:	Long – more than ¹³/₁₆ inch.
Flowering Habit:	Medium flowering and fruiting.
Hardiness:	Zones 6 to 8.
Plant Use:	Specimen.

Registration: Gagnep. in *Bull. Soc. Bot. Fr.*, 68:482, 1921; in *Lecte. Fl. Gen. Indo-Chine*, 5:661, f.77, 10, 78, 1-4, 1927; Hatusima in *Journ. Dept. Agr. Kyushu Univ.*, 6(6):288, 1942; *Yunnan Journal of Plants*, 1:142, f.35, 1-3, 1977; Zhen Mian, Min Tianly in *Flora Reipublicae Popularis Sinicae, Tomus* 45(1), Science Press, 1980.
History: Found in Guangxi (the region of Tiane, Hechi, Lingyun and Fengshan) and Yunnan (Funing); grows on slopes, beside streams, under forest cover. Distributed through Vietnam and Laos.
Bibliography:
The Boxwood Bulletin, Vol. 30(3):52-53, Jan. 1991.
Known Locations: Not documented
Additional Information: Not documented

Buxus linearifolia – Species

Size (25 yrs.):	Small – 2½ to 3 feet in height.
Natural Form:	Described as a shrub.
Annual Growth:	Medium – 1½ to 2 inches in height.
Leaf Color:	Not observed.
Leaf Shape:	Linear; obtuse tip, commonly retuse; narrow cuneate base.
Leaf Size:	Large – ½ to 1 inch long and ⅛ to 3/16 inch wide.
Leaf Surface:	Glabrous and smooth.
Internodal Length:	Short – 3/16 to ¼ inch.
Flowering Habit:	Medium flowering and fruiting.
Hardiness:	Zones 6 to 8.
Plant Use:	Specimen.

Registration: Zheng Mian and Min Tianly in *Flora Reipublicae Popularis Sinicae*, *Tomus* 45(1), Science Press 1980.
History: Found in the Shiwan da mountains south of Sidong in Guangxi.
Bibliography:
The Boxwood Bulletin, Vol. 31(1):17, July 1991.
Known Locations: Not documented
Additional Information: Not documented

Buxus megistophylla – Species

Size (25 yrs.):	Large – up to 6¾ feet in height.
Natural Form:	Described as shrub or small tree.
Annual Growth:	Medium – 3 to 3½ inches in height.
Leaf Color:	Not observed.
Leaf Shape:	Elliptical or oblong-lanceolate to lanceolate; acute tip; cuneate base, margin curves downward.
Leaf Size:	Large – 4 to 8 cm long and 1.5 to 3 cm wide.
Leaf Surface:	Glossy and shiny.
Internodal Length:	Not observed.
Flowering Habit:	Medium flowering and fruiting.
Hardiness:	Zones 6 to 8.
Plant Use:	Specimen.

General Comment: Illustration not available.
Registration: Levl. Fl. Kouy-Tcheou 160, 1914; Gagnep. in *Lecte. Fl. Gen. Indo-Chine*, 5:661, 1927; Rehd. in *Journ. Arn. Arb.*, 14:236, 1933; Hatusima in *Journ. Dept. Agr. Kyushu Univ.*, 6(6):284, f.5, 1942; Zheng Mian, Min Tianly in *Flora Reipublicae Popularis Sinicae, Tomus* 45(1), Science Press, 1980.
History: Found in southwest Guizhou (Zhenning, Loudian), northeast Guangxi (Lingui, Guanyang), northwestern Guangdong (the Lian County region), south Hunnan (Yizhang) and southern Jiangxi (Anyuan, Huichang); grows in hills, valleys, river banks or wooded slopes, altitude 500-1400 m. Type specimen collected from Loudian in Guizhou, and Huangguoshu between Zhenning and Guangling.
Bibliography:
The Boxwood Bulletin, Vol. 30(4):69, April 1991.
Known Locations: Not documented
Additional Information: Not documented

Buxus microphylla – Species

General Comment: Variation exists within this species; and it is best represented by the range exhibited by the varieties and cultivars of this species.
Registration: P.E. Siebold and J.C. Zuccarini in *Abhandl. Math. Phys. Konigl. Akad. Wissensch. Munch.*, 4(2):142, 1845.
History: Dr. H.T. Skinner of the Arnold Arboretum and former director of the U.S. National Arboretum wrote in 1967:
"Basic or first named *Buxus microphylla* Siebold and Zucc. (Himetsuge is its Japanese name) is usually described as a compact rounded shrub to 2 or 3 feet high with square, glabrous shoots bearing round-ended obovate leaves, 1/2 to 3/4 inch long, flowers in terminal clusters. It has been cultivated in Japan at least since 1450 and was introduced to western gardens around 1860. However, according to Ohwi in his *Flora of Japan*, 1965, this boxwood is not known in the wild state and its origin is somewhat of a mystery. It could have been lost in the wild; it could have been carried to Japan from Korea or China; or it could even, possibly, be a hybrid or segregate from the much larger Japanese boxwood, which seems comparatively adept at producing diminutive or aberrant forms. As in the case of some other garden plants it seems unsatisfactory to be dealing with the type of Species which cannot be definitely pegged down."
Bibliography:
Bender, S. *Southern Living*, Nov. 1987.
Bush-Brown, J. and L. *America's Garden Book*, Rev. New York Botanical Garden. 1980.
Dirr, M.A. *Manual of Woody Landscape Plants*, Third Ed., 1983.
Krussman, G. *Manual of Cultivated Broad-leaved Trees and Shrubs*, Vol. I, A-D, 1984.
Taylor, N. *Taylor's Encyclopedia of Gardening*, 4th Ed.
Wright, M. *The Complete Book of Gardening*, 1984.
Wyman, D. *Wyman's Gardening Encyclopedia*, New Expanded 2nd Ed., 1986.
The Boxwood Bulletin, Vol. 2(3):8, Jan. 1963/2(4):43, April 1963/6(3):38, Jan. 1967/ 7(1):12, July 1967/14(1):13, July 1974/16(1):12,14, July 1976/17(2):25, Oct. 1977/ 20(4):80, April 1981/21(3):45, Jan. 1982.
Known Locations: Available in the commercial nursery trade but not uniformly described.

Buxus microphylla variety *japonica* - Variety

Size (25 yrs.):	Medium – 5 to 6 feet high and 7 to 8 feet wide. A 53-year-old specimen measures 6½ feet high and 12 feet wide.
Natural Form:	Mounded, semi-globular.
Annual Growth:	Fast – 3 to 3½ inches in height and 4 to 4½ inches in width.
Leaf Color:	Medium yellow-green.
Leaf Shape:	Obovate; obtuse tip, occasionally retuse; cuneate base.
Leaf Size:	Large – ⅝ to ⅞ inch long and ⅜ to 7/16 inch wide.
Leaf Surface:	Glossy and smooth.
Internodal Length:	Long – ⅜ to ½ inch.
Flowering Habit:	Floriferous; heavy fruiting.
Hardiness:	Zones 5 to 10.
Plant Use:	Specimen, hedging, grouping for background and area separations.

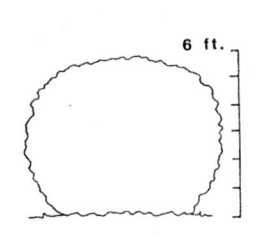

Registration: (J. Mueller, Arg.) Rehder and Wilson, *Plantae Wilsonianae*, 2(1):168, 1914.
History: Indigenous to the mountains of Honshu, Shikoku and Kyushu, Japan. Came to North America about 1860.
Bibliography:
Bailey, L.H. *Hortus Third*, 1976.
Bender, S. *Southern Living*, Nov. 1987.
Bush-Brown, J. and L. *America's Garden Book*, Rev., New York Botanical Garden, 1980.
Dallimore, W. *Holly, Yew and Box*, 1908.
Dirr, M.A. *Manual of Woody Landscape Plants*, 4th Ed., 1990.
Everett, T.H. *The New York Botanical Garden Ill. Encyc. of Hort.*, Vol. 2, 1981.
Flint, H.L. *Horticulture*, March 1987.
Gamble, M.A. *Flower and Garden*, March 1988.
Krussman, G. *Manual of Cultivated Broad-leaved Trees and Shrubs*, Vol. I, A-D, 1986.
Missouri Botanical Garden Bulletin, Vol. LXXVI (3):8-10, May-June 1988.
Stone, B.M. *American Horticulturist*, June 1987.
Taylor, N. *Taylor's Encyclopedia of Gardening*, 4th Ed.
The Boxwood Society of the Midwest Bulletin, April 1986.
Wright, M. *The Complete Handbook of Garden Plants*, 1984.

The Boxwood Bulletin, Vol. 2(3):28, Jan. 1963/2(4):43, April 1963/6(3):38, Jan. 1967/7(1):12-14, July 1967/8(4):54, April 1969/14(1):13, July 1974/16(1):8-9,12,14, July 1976/17(3):39, Jan. 1978/20(3):42,44, Jan. 1981/20(4):80, April 1981/21(3):45, Jan. 1982/21(4):7, April 1982/23(1):19, July 1983/27(3):58,64, Jan. 1988.

Known Locations: Arnold Arboretum, Brooklyn Botanic Garden; Buzzards Bay Garden Club; College of William and Mary; Royal Botanic Gardens, Edinburgh; George Landis Arboretum; Hillier's Arboretum; Royal Botanic Gardens, Kew; Missouri Botanical Garden; Morton Arboretum; Secrest Arboretum; U.S. National Arboretum; State Arboretum of Virginia.

Additional Information:
Culture and Care: Prefers dappled shade but will tolerate siting in some direct sun. Breaks dormancy early.
Pests and Diseases: Resistant to leaf miner and psyllid.

Buxus mollicula – Species

Size (25 yrs.):	Large – 8½ to 10 feet in height.
Natural Form:	Described as a shrub.
Annual Growth:	Fast – 4 to 5 inches in height.
Leaf Color:	Not observed.
Leaf Shape:	Ovate; acute tip; cuneate base.
Leaf Size:	Medium – 3/4 to 13/16 inch long and 1/4 to 5/16 inch wide.
Leaf Surface:	Glabrous and smooth.
Internodal Length:	Medium – 3/8 to 1/2 inch.
Flowering Habit:	Medium flowering and fruiting.
Hardiness:	Zones 6 to 8.
Plant Use:	Specimen.

Registration: W.W. Smith in *Not. Roy. Bot. Gard. Edinb.*, 10:16, 1917; Hand.-Mazz. Symb. Sin. 7:236, 1931; Hatusima in *Journ. Dept. Agr. Kyushu Univ.*, 6(6):294, f.10, c-g, 1942; *Yunnan Journal of Plants*, 1:144, figure 35, 8-11, 1977. *Buxus wallichiana* Baill. var. *velutina* Franch. Pl. Delav. 136, 1889; Zheng Mian, Min Tianly in *Flora Reipublicae Popularis Sinicae, Tomus* 45(1), Science Press, 1980.
History: Found in northwestern Yunnan (Ninglang, Yongsheng, Binchuan, Lijiang) in the forests of the Jinsha River valley; altitude below 1780 m. Type specimen collected in Lijiang, Yunnan.
Bibliography:
The Boxwood Bulletin, Vol. 30(4):69, April 1991.
Known Locations: Not documented
Additional Information: Not documented

Buxus myrica – Species

Size (25 yrs.):	Large – more than 6 feet in height.
Natural Form:	Described as a shrub.
Annual Growth:	Fast.
Leaf Color:	Not observed.
Leaf Shape:	Lanceolate; acute tip; cuneate base.
Leaf Size:	Large – 1³/₈ to 1¹/₂ inches long and ¹/₄ to ³/₈ inch wide.
Leaf Surface:	Glabrous and smooth.
Internodal Length:	Medium – ¹/₂ to ³/₄ inch.
Flowering Habit:	Medium flowering and fruiting.
Hardiness:	Zones 6 to 9.
Plant Use:	Specimen.

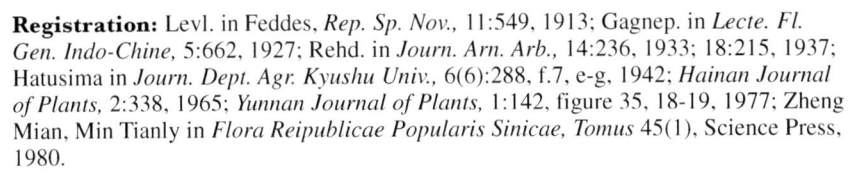

Registration: Levl. in Feddes, *Rep. Sp. Nov.*, 11:549, 1913; Gagnep. in *Lecte. Fl. Gen. Indo-Chine*, 5:662, 1927; Rehd. in *Journ. Arn. Arb.*, 14:236, 1933; 18:215, 1937; Hatusima in *Journ. Dept. Agr. Kyushu Univ.*, 6(6):288, f.7, e-g, 1942; *Hainan Journal of Plants*, 2:338, 1965; *Yunnan Journal of Plants*, 1:142, figure 35, 18-19, 1977; Zheng Mian, Min Tianly in *Flora Reipublicae Popularis Sinicae, Tomus* 45(1), Science Press, 1980.

History: Found in sub-central Guizhou (between Pingba, Luodian and Wangmo), western and northern Guangxi, northwestern Yunnan (Yilang), Guangdong (Hainan Island), Hunnan and Sichuan; grows by streams, on slopes and in woods, altitude 250-2000 m. Distributed through Vietnam. Type specimens collected from Pingba and Luodian in Guizhou.

Bibliography:
The Boxwood Bulletin, Vol. 30(4):67 April 1991.
Known Locations: Not documented
Additional Information: Not documented

Buxus pubiramea – Species

Size (25 yrs.):	Large – 9 to 10 feet in height.
Natural Form:	Described as a shrub.
Annual Growth:	Fast.
Leaf Color:	Not observed.
Leaf Shape:	Lanceolate; obtuse tip, often retuse; cuneate base narrow and pointed, margin chondroid.
Leaf Size:	Large – more than $^{13}/_{16}$ inch in length and $^{1}/_{2}$ inch in width.
Leaf Surface:	Glabrous and smooth.
Internodal Length:	Not observed.
Flowering Habit:	Not observed.
Hardiness:	Zones 6 to 8.
Plant Use:	Specimen.

Registration: Merrill et Chun in *Sunyats.*, 5:104, 1940; *Hainan Journal of Plants*, 2:339, 1956; Zheng Mian, Min Tianly in *Flora Reipublicae Popularis Sinicae, Tomus 45(1)*, Science Press, 1980.
History: Found in Guangdong (Baoting in Hainan Island); grows on wooded slopes, altitude 650 m. Type specimen collected in Baoting in Hainan Island.
Bibliography:
The Boxwood Bulletin, Vol. 30(3):51-52, Jan. 1991.
Known Locations: Not documented
Additional Information: Not documented

Buxus rugulosa – Species

Size (25 yrs.):	Medium – 3½ to 6 feet in height.
Natural Form:	Described as a shrub.
Annual Growth:	Medium – 1½ to 2 inches in height.
Leaf Color:	Not observed.
Leaf Shape:	Lanceolate; rounded tip; cuneate base.
Leaf Size:	Medium – ⅝ to ⅞ inch long and 3/16 to ¼ inch wide.
Leaf Surface:	Glabrous and smooth.
Internodal Length:	Medium – ⅜ to ½ inch.
Flowering Habit:	Sparse flowering and fruiting.
Hardiness:	Zones 6 to 8.
Plant Use:	Specimen.

Registration: Hatusima in *Journ. Agr. Kyushu Univ.*, 6(6):303, f.15, a-b, Pl. 22(7), f.2, 1942; *Flora Yunnanensis*, 1:145, figure 35, 12-14, 1937; – *B. microphylla* Siebold et Zucc. var. *platyphylla* (Schneid.) Hand.-Mazz. Symb. Sin. 7:237, 1931, excl. syn.; Zheng Mian, Min Tianly in *Flora Reipublicae Popularis Sinicae, Tomus* 45(1), Science Press, 1980.

History: Found in northwestern Yunnan (Deqin and Lijiang to Weishan) and Sichuan (Maerkang and Jinzhuan); grows beside streams and in mountain bush-wood, altitude 1900-3500 m. Type specimen collected at an altitude of 3000 m in the vicinity of Lijiang.

Bibliography:
The Boxwood Bulletin, Vol. 30(4):72-73, April 1991.

Known Locations: Not documented
Additional Information: Not documented

Selected Species, Varieties, Cultivars

Buxus sempervirens – Species

Size (25 yrs.):	Highly variable, most are medium to large.
Natural Form:	Highly variable.
Annual Growth:	Highly variable.
Leaf Color:	Variable.
Leaf Shape:	Variable.
Leaf Size:	Variable, ranging from $1/4$ inch to as much as nearly 2 inches long and from $1/8$ inch to as much as $1 1/2$ inches wide.
Leaf Surface:	Variable.
Internodal Length:	Variable.
Flowering Habit:	Variable.
Hardiness:	Variable, ranging from Zone 5 through Zone 10.
Plant Use:	Many applications within the landscape setting.

General Comment: The great variation within this species is best represented by the range exhibited by the varieties and cultivars of this species.
Registration: Linnaeus, *Species Plantarum,* 983. 1753.
History: All evidence indicates that the species described by Linnaeus in the 1700s has been lost in cultivation and the only identifiable remains are located in the herbarium collection at the Royal Bontanic Gardens, Kew. The species is native to the limestone formations of a large cross section of Southern Europe. From the mountainous areas of southern France, the Pyrenees and Portugal it flows over the hills of northern Italy and into Yugoslavia, the remaining Balkan states and Greece. It spreads around the Black Sea to the Caucasas Mountains, with layers scattered out in Algeria, Germany, Belgium and England. The proliferation of open-pollinated seedlings through the centuries has widely increased this species' adaptability to both soil characteristics and climatic conditions. Erroneously called American box in the trade, although it is not native to the western hemisphere.

Buxus sinica – Species

Buxus sinica is a species whose natural habitat is Korea. It has had a checkered career in taxonomic nomenclature. One may find it described as *B. microphylla* variety *koreana, B. microphylla* variety *insularis, B. sinica* subsp. *sinica* variety *insularis, B. microphylla* variety *sinica*, or just plain Korean box.

Since this book is not intended to be a treatise on boxwood taxonomy, we have elected to use the name *Buxus sinica* variety *insularis* for the plant commonly known as Korean boxwood. Although many other forms of *Buxus sinica* grow in Asia and have been described, none have been introduced into horticultural use. Therefore, treatment of this species is limited to the variety *insularis*.

This is consistent with the work of Dr. Peter Goldblatt of the Missouri Botanical Garden who wrote:

"The second most important group of boxwood (after *Buxus sempervirens*) are the cultivars of two East Asian species, until recently regarded as all belonging to a single species *B. microphylla* Siebold et Zucc. (Hatusima 1942). However, in 1980 a treatment of *Buxaceae* was published by Cheng Mien in *Flora of the Peoples Republic of China* in which the name *B. microphylla* is restricted to plants from Japan. According to Cheng, cultivars of boxwoods of Korean and Chinese origin all belong to *sinica* subsp. *sinica*.

The important small-leaved and very hardy Korean boxwood *Buxus koreana* of North American gardens is *B. sinica* subsp. *sinica* variety *insularis* (Nakai) M. Cheng (known botanically for the past 40 years as *B. microphylla* variety *insularis* (Nakai) Hatus.). The varietal name var. *koreana* is predated by two years by var. *insularis*, hence the confusion in having to use the latter name, var. *insularis*, for the Korean boxwoods....There are now several named cultivars of Korean boxwood and for nomenclature purposes they should best be identified as, for example, *Buxus sinica* var. *insularis* 'Wintergreen.' "

(*The Boxwood Society of the Midwest Bulletin*, Vol. XI(1):7, Jan. 1987).

Buxus sinica variety *insularis* - Variety

Size (25 yrs.):	Medium – 5 to 5^1/$_2$ feet high and 7 to 7^1/$_2$ feet wide.
Natural Form:	Mounded.
Annual Growth:	Slow – 1 to 1^1/$_2$ inches in height and 1^1/$_2$ to 2 inches in width.
Leaf Color:	Light yellow-green.
Leaf Shape:	Ovate; obtuse tip; cuneate base.
Leaf Size:	Small – ^{11}/$_{16}$ to ^{3}/$_4$ inch long and ^{1}/$_4$ to ^{5}/$_{16}$ inch wide.
Leaf Surface:	Glabrous and smooth.
Internodal Length:	Long – ^{1}/$_2$ to ^{5}/$_8$ inch.
Flowering Habit:	Sparse flowering and sparse fruiting.
Hardiness:	Zones 5 to 8.
Plant Use:	Specimen, foundation planting, grouping for background and area separations, hedgings.

Registration: Nakai in *Botanical Magazine of Tokyo*, 36:63, 1922.
History: E.H. Wilson of the Arnold Arboretum brought the first cuttings from Korea to the United States in 1919. They were accessioned as No. 11323. During the ensuing years this variety has been classified as *Buxus microphylla* variety *koreana* and *Buxus microphylla* variety *insularis*.
Bibliography:
Bailey, L.H. *Hortus Third*, 1976.
Bush-Brown, J. and L. *America's Garden Book*, Rev., New York Botanical Garden, 1980.
Dirr, M.A. *Manual of Woody Landscape Plants*, 4th Ed., 1990.
Flint, H.L. *Horticulture,* March 1987.
Gamble, M.A. *Flower and Garden,* March 1988.
Krussman, G. *Manual of Cultivated Broad-leaved Trees and Shrubs*, Vol. I, A-D, 1986.
Taylor, N. *Taylor's Encyclopedia of Gardening*, 4th Ed.
Wright, M. *The Complete Handbook of Garden Plants*, 1984.
Wyman, D. *Wyman's Gardening Encyclopedia*, New Expanded 2nd Ed., 1986.
Missouri Botanical Garden Bulletin, Vol. LXXVI(3):8-10, May-June 1988.
The Boxwood Society of the Midwest Bulletin, April 1986.
The Boxwood Bulletin, Vol. 1(2):cover, ifc, Jan. 1962/2(4):44, April 1963/4(2):24, Oct. 1964/5(3):40-41, Jan. 1966/6(1):6, July 1966/7(1):12-15, July 1967/7(2):26-27, Oct. 1967/8(4):50-54, April 1969/14(1):13, July 1974/17(2):25, Oct. 1977/ 17(3):39, Jan. 1978/19(2):31, Oct. 1979/20(3):43,45, Jan. 1981/20(4):80, April 1981/21(3):45, Jan. 1982/21(4):62, April 1982/22(4):71, April 1982/23(1):18-19, July 1983.

Known Locations: Arnold Arboretum, Brooklyn Botanic Garden, Buzzards Bay Garden Club, College of William and Mary, Royal Botanic Gardens, Edinburgh, George Landis Arboretum, Hillier's Arboretum, Royal Botanic Gardens, Kew, Missouri Botanical Garden, Secrest Arboretum, Tennessee Botanical Gardens, U.S. National Arboretum, State Arboretum of Virginia.

Additional Information:
Culture and Care: Demonstrates no special cultural requirements.
Pests and Diseases: Indicates resistance to leaf miner, psyllid and mites in the more humid climates; no serious diseases.
Available in the commercial nursery trade.

Selected Species, Varieties, Cultivars

Buxus stenophylla – Species

Size (25 yrs.):	Medium – 5 feet in height.
Natural Form:	Described as a shrub.
Annual Growth:	Medium – $2^1/_2$ to 3 inches in height.
Leaf Color:	Dark green.
Leaf Shape:	Obovate; ovate tip, often retuse; cuneate base.
Leaf Size:	Medium – $7/_{16}$ to $3/_4$ inch long, $3/_{16}$ to $3/_8$ inch wide.
Leaf Surface:	Glabrous and smooth, midrib prominent on both sides.
Internodal Length:	Medium – $3/_{16}$ to $1/_4$ inch.
Flowering Habit:	Medium flowering and fruiting.
Hardiness:	Zones 6 to 8.
Plant Use:	Specimen.

Registration: Hance in *Journ. Bot. Brit. et For.,* 6:331, 1868; in *Journ. Linn. Soc. Bot.,* 13:124, 1873 in nota; Muell.-Arg. in *DC. Prodr.,* 16(1):20, 1869; Rehd. et Wils. in *Sarg. Pl. Wils.,* 2:169, 1914; Hatusima in *Journ. Dept. Agr. Kyushu Univ.,* 6(6):318, f.22, 1942 – *Buxus sempervirens* auct. non L.; Hemsl. in *Journ. Linn. Soc.,* 26:418, 1894, p.p. quoad Syn; Zheng Mian, Min Tianly in *Flora Reipublicae Popularis Sinicae, Tomus* 45(1), Science Press, 1980.

History: Found in southeastern Fujian (Anxi), Guandong (Lian county and vicinity of Guangzhou) and Guizhou (Qingxi); grows on river banks or in forests. Type specimen collected from Anxi in Fujian.

Bibliography:
The Boxwood Bulletin, Vol. 30(4):73-74, April 1991.

Known Locations: Not documented

Additional Information: Not documented

Buxus wallichiana – Species

Size (25 yrs.):	Large – 8 to 9 feet high and 6 to 7 feet wide.
Natural Form:	Pyramidal and somewhat loose and open.
Annual Growth:	Fast – $3^1/_2$ to 4 inches in height and $2^1/_2$ to 3 inches in width.
Leaf Color:	Dark green.
Leaf Shape:	Elliptic to lanceolate; acute tip; cuneate base.
Leaf Size:	Large – $1^1/_4$ to $1^1/_2$ inches long and $^3/_8$ to $^1/_2$ inch wide.
Leaf Surface:	Glabrous and smooth.
Internodal Length:	Medium – $^3/_8$ to $^1/_2$ inch.
Flowering Habit:	Sparse flowering and sparse fruiting.
Hardiness:	Zones 7 to 9.
Plant Use:	Specimen.

Registration: H. Baillon in *Monographie des Buxacées et des Stylocérées,* 63-64, 1859.
History: Krussman states: "small shrubs, some 1 m high in culture but tree-like in wild, young twigs remaining densely pubescent; leaves lanceolate to ovate-lanceolate, usually 4-5 cm long, 7-10 cm wide, widest in middle, dark green, petiole and midrib pubescent; flowers in dense, axillary clusters, anthers yellow, April BS1:464, India; NW Himalaya, 1850, 28." Dr. P. Goldblatt has written that this is a Himalayan species, probably introduced as a curiosity. It reaches a height of 6 m and is a useful timber tree in India. It seldom reaches this height in cultivation. Dallimore reports that it is a strong grower but difficult to propagate. Beckett describes it as a Himalayan boxwood, leaves 2.5 to 6 cm long, lance-shaped and glossy. About 2 m in cultivation.
Bibliography:
Becket, K.A. *The Complete Book of Evergreens,* 1981.
The Boxwood Bulletin, Vol. 16(1):July 1976/26(2):26-27, Oct. 1986/27(4):80, April 1988.
Known Locations: Royal Botanic Gardens, Edinburgh, Hillier's Arboretum, Royal Botanic Gardens, Kew.

Buxus harlandii 'Richard'

Size (25 yrs.): Medium – 4½ to 5 feet high and 4½ to 5 feet wide.
Natural Form: Unusual – upright, loose and open habit.
Annual Growth: Medium – 2 to 2¼ inches high and 2 to 2¼ inches wide.
Leaf Color: Medium green.
Leaf Shape: Obovate; obtuse and retuse tip; cuneate base.
Leaf Size: Large – 1 to 1⅛ inches long and ½ to ¾ inch wide.
Leaf Surface: Glossy, smooth and shiny.
Internodal Length: Long – ½ to ¾ inch.
Flowering Habit: Not observed.
Hardiness: Zones 7(protected) to 10.
Plant Use: Specimen.

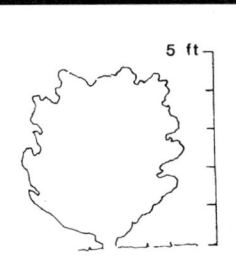

Registration: J.T. Baldwin in *The Boxwood Bulletin,* 2(4):44, April 1963.
History: Introduced by Cassadaban Nurseries, Abita Springs, La., and Gulf Stream Nurseries, Wachapreague, Va.
Bibliography:
The Boxwood Bulletin, Vol. 2(4):44, April 1963.
Flint, H.L. *Horticulture,* March 1987.
Known Locations: College of William and Mary, Hillier's Arboretum, State Arboretum of Virginia.
Additional Information:
Culture and Care: Prefers dappled shade and to be out of the winter winds. Breaks dormancy early, prone to dieback from late freezes. Dieback not injurious, just ugly. Pruning will eliminate the problem.
Pests and Diseases: Resistance to leaf miner, spider mite and psyllid.
Propagation: Bruise the stem tissue, use an IBA powder dip.

Buxus microphylla 'Compacta'

Size (25 yrs.):	Dwarf – 10 to 12 inches high; 25 to 28 inches wide. A 47-year-old specimen measures 12 inches high and 48 inches wide.
Natural Form:	Mounded – twiggy and uneven, sports frequently; is the mother clone of several named cultivars.
Annual Growth:	Slow – less than 1/2 inch high and 3/4 inch wide.
Leaf Color:	Light yellow-green.
Leaf Shape:	Ovate; obtuse tip, some slightly retuse; cuneate base.
Leaf Size:	Small – 5/16 to 3/8 inch long and 3/16 to 1/4 inch wide.
Leaf Surface:	Glossy, smooth and shiny.
Internodal Length:	Short – 1/8 to 3/16 inch.
Flowering Habit:	Not observed.
Hardiness:	Zones 5(protected) to 9.
Plant Use:	Specimen, container plant, edgings, bonsai.

Registration: D. Wyman in *American Nurseryman*, 107(7):50, 1963; Henry Hohman, Kingsville Nurseries, Kingsville, Md., in *The Boxwood Bulletin*, 4(3):26, Jan. 1965. Invalid classification: *Buxus microphylla* variety *compacta* 'Kingsville Dwarf.'
History: Originated with Sam Appleby in Baltimore, Md., in 1912. Upon his death in the 1920s it was acquired by Henry Hohman of Kingsville Nurseries, Kingsville, Md., and was commercially released in 1937.
Bibliography:
Bender, S. *Southern Living*, Nov. 1987.
Dirr, M.A. *Manual of Woody Landscape Plants*, 4th Ed., 1990.
Gamble, M.A. *Flower and Garden*, March 1988.
Krussman, G. *Manual of Cultivated Broad-leaved Trees and Shrubs*, Vol. I, A-D, 1984.
Wright, M. *The Complete Book of Garden Plants*, 1984.
The Boxwood Bulletin, Vol. 2(4):43, April 1963/5(1):7, July 1965/8(4):54, April 1969 14(1):12, July 1974/20(3):42,45, Jan. 1981/21(3):45, Jan. 1982/21(4):61, April 1982/23(3):63, Jan. 1984/27(2):37, Oct. 1987/27(3):65, Jan. 1988.
Known Locations: Brooklyn Botanic Garden, Buzzards Bay Garden Club, College of William and Mary, Dixon Gallery and Garden, George Landis Arboretum, Hillier's Arboretum, Longwood Gardens, Morton Arboretum, Royal Horticultural Society, Secrest Arboretum, Washington Park Arboretum, U.S. National Arboretum, State Arboretum of Virginia.

Additional Information:
Culture and Care: Prefers dappled shade. Pruning of sports (mutations) improves appearance.
Pests and Diseases: Resistance to leaf miner, psyllid and spider mites.
Propagation: Difficult and quite slow to root. Take cuttings when spring growth has hardened and include a portion of the heel from the previous year's growth; bruise the stem tissue or slit each side of the stem; use an IBA powder dip.

Buxus microphylla 'Creepy'

Size (25 yrs.):	Dwarf – 1 to 1½ feet high; 1½ to 2 feet wide.
Natural Form:	Mounded – semi-globular.
Annual Growth:	Slow – ½ to ⅝ inch high and ⅝ to ¾ inch wide.
Leaf Color:	Medium green.
Leaf Shape:	Elliptic; obtuse tip; cuneate base.
Leaf Size:	Small – ½ to ¾ inch long and 3/16 to ¼ inch wide.
Leaf Surface:	Glossy and smooth.
Internodal Length:	Short – 3/16 to ¼ inch.
Flowering Habit:	Not observed.
Hardiness:	Zones 6 to 8.
Plant Use:	Specimen, edging, grouping for background and area separations.

Registration: Not registered.
History: Discovered and named by Lorenz C. Frank of Split Rock Nurseries, Paramus, N.J. Originated from a mutation of *Buxus microphylla* 'Compacta.'
Bibliography:
The Boxwood Bulletin, Vol. 28(3):44,55, Jan. 1989/28(4):74, April 1989.
Known Locations: State Arboretum of Virginia.
Additional Information:
Culture and Care: Prefers dappled shade.
Pests and Diseases: Resistance to leaf miner, psyllid and mites.
Propagation: Bruise the stem or slit each side; use an IBA powder dip.

Selected Species, Varieties, Cultivars

Buxus microphylla 'Curly Locks'

Size (25 yrs.):	Medium – 4½ to 5 feet high and 3½ to 4 feet wide. A 40-year-old specimen measures 8 feet high and 9 feet wide.
Natural Form:	Unusual – open upright habit, twisted and irregular branches.
Annual Growth:	Medium – 2 to 2½ inches high and 1½ to 2 inches wide.
Leaf Color:	Light yellow-green.
Leaf Shape:	Obovate, moderately revolute; obtuse tip; cuneate base.
Leaf Size:	Small – ³/₈ to ⁵/₁₆ inch long and ¼ to ⁵/₁₆ inch wide.
Leaf Surface:	Glabrous and smooth.
Internodal Length:	Short – ¼ to ⁵/₁₆ inch.
Flowering Habit:	Floriferous; heavy fruiting, highly abortive.
Hardiness:	Zones 5(protected) to 8.
Plant Use:	Specimen; espalier.

Registration: D. Wyman in *American Nurseryman*, 107(7):50, 1963.
History: Originated with Henry Hohman of Kingsville Nurseries, Kingsville, Md., in the 1930s as a mutation from *Buxus microphylla* 'Compacta.'
Bibliography:
Dirr, M.A. *Manual of Woody Landscape Plants*, 4th Ed., 1990.
Flint, H.L. *Horticulture,* March 1987.
Wyman, D. *Wyman's Gardening Encyclopedia*, New Expanded 2nd Ed., 1986.
The Boxwood Society of the Midwest Bulletin, April 1986.
The Boxwood Bulletin, Vol. 14(1):12, July 1974/16(1):14, July 1976/17(2):25, Oct. 1977/18(2):43, Oct. 1978/18(3):54, Jan. 1979/20(3):43-44,46, Jan. 1981/20(4):80, April 1981/21(4):61, April 1982/27(3):Jan. 1988.
Known Locations: Arnold Arboretum, Buzzards Bay Garden Club, College of William and Mary, Hillier's Arboretum, Longwood Gardens, Missouri Botanical Garden, Washington Park Arboretum, U.S. National Arboretum.
Additional Information:
Culture and Care: Prefers dappled shade.
Pests and Diseases: Resistance to leaf miner, psyllid and mites.
Propagation: Somewhat difficult; quite slow to root. Take cuttings when spring growth has hardened and include a portion of the heel from the previous year's growth; bruise the stem tissue or slit each side of the stem; use an IBA powder dip.

Buxus microphylla 'Grace Hendrick Phillips'

Size (25 yrs.):	Dwarf – 1 to 1½ feet high and 3 to 3½ feet wide.
Natural Form:	Mounded, semi-globular, twiggy, brittle.
Annual Growth:	Slow – ½ to ¾ inch high and 1½ to 2 inches wide.
Leaf Color:	Medium yellow-green.
Leaf Shape:	Obovate to lanceolate; obtuse tip; cuneate base.
Leaf Size:	Small – 7/16 to 11/16 inch long and ¼ to 5/16 inch wide.
Leaf Surface:	Glossy and smooth.
Internodal Length:	Short – 3/16 to ¼ inch.
Flowering Habit:	Not observed.
Hardiness:	Zones 6 to 8.
Plant Use:	Specimen, edging, grouping for background and area separations, bonsai.

Registration: Henry Hohman in *The Boxwood Bulletin,* Vol. 7(1):1, July 1967.
History: Discovered as a sport of *Buxus microphylla* 'Compacta,' propagated and commercially released by Henry Hohman of Kingsville Nurseries, Kingsville, Md. Named for the wife of Admiral Neill Phillips, former president of The American Boxwood Society.
Bibliography:
The Boxwood Bulletin, Vol. 7(1):1, July 1967.
Known Locations: College of William and Mary, Washington Park Arboretum, U.S. National Arboretum, State Arboretum of Virginia.
Additional Information:
Culture and Care: Requires dappled shade.
Pests and Diseases: Resistance to leaf miner, psyllid and mites.
Propagation: Difficult and quite slow to root. Take cuttings when spring growth has hardened and include a portion of the heel from the previous year's growth; bruise the stem tissue or slit each side of the stem; use an IBA powder dip.

Buxus microphylla 'Green Pillow'

Size (25 yrs.):	Medium – 3 to 3½ feet high and 3½ to 4 feet wide.
Natural Form:	Mounded – uneven, twiggy and brittle.
Annual Growth:	Slow – ¾ to 1 inch high and 1 to 1¼ inches wide.
Leaf Color:	Dark green.
Leaf Shape:	Ovate; obtuse tip, some retuse; cuneate base.
Leaf Size:	Small – ⅜ to 7/16 inch long and 3/16 to ¼ inch wide.
Leaf Surface:	Matte – somewhat dull.
Internodal Length:	Short – 3/16 to ¼ inch.
Flowering Habit:	Not observed.
Hardiness:	Zones 5(protected) to 8.
Plant Use:	Specimen, grouping for background and area separations, edgings, hedgings, bonsai.

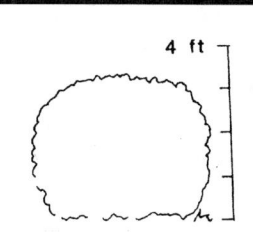

Registration: O'Connor in *Baileya* 1:114, 1963.
History: Originated with Sam Appleby of Baltimore, Md., in the early 1900s. Transferred to Henry Hohman of Kingsville Nurseries, Kingsville, Md., in the 1920s and carried as "Kingsville 1A." Named by Mr. Hohman and commercially released in 1953. Many were planted at the White House in 1962.
Bibliography:
Dirr, M.A. *Manual of Woody Landscape Plants*, 4th Ed., 1990.
Flint, H.L. *Horticulture,* March 1987.
Krussman, G. *Manual of Cultivated Broad-leaved Trees and Shrubs*, Vol. I, A-D, 1984.
Wright, M. *The Complete Handbook of Garden Plants*, 1984.
Wyman, D. *Wyman's Gardening Encyclopedia*, New Expanded 2nd Ed., 1986.
The Boxwood Society of the Midwest Bulletin, April 1986.
The Boxwood Bulletin, Vol. 15(3):41-42, Jan. 1976/19(2):26, Oct. 1979/20(3):42,44, Jan. 1981/21(4):61, April 1982/23(3):63, Jan. 1984.
Known Locations: Arnold Arboretum, Brooklyn Botanic Garden, Buzzards Bay Garden Club, College of William and Mary, Hillier's Arboretum, Missouri Botanical Garden, Secrest Arboretum, Washington Park Arboretum, U.S. National Arboretum, State Arboretum of Virginia.
Additional Information:
Culture and Care: Prefers dappled shade.
Pests and Diseases: Resistance to leaf miner, psyllid and mites.

Buxus microphylla 'Helen Whiting'

Size (25 yrs.):	Medium – 4 to $4^1/_2$ feet high and 5 to $5^1/_2$ feet wide.
Natural Form:	Mounded - uneven and twiggy.
Annual Growth:	Slow – $1^1/_2$ to 2 inches high and 2 to $2^1/_2$ inches wide.
Leaf Color:	Light green.
Leaf Shape:	Obovate to lanceolate; obtuse tip, some retuse; cuneate base.
Leaf Size:	Small – $^1/_2$ to $^{11}/_{16}$ inch long and $^1/_4$ to $^3/_8$ inch wide.
Leaf Surface:	Glossy and smooth.
Internodal Length:	Short – $^1/_4$ to $^3/_8$ inch.
Flowering Habit:	Not observed.
Hardiness:	Zones 6 to 8.
Plant Use:	Specimen, grouping for background and area separations, edgings, hedgings, bonsai.

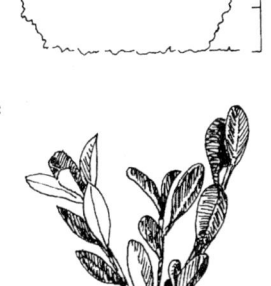

Registration: Dr. J.T. Baldwin, Jr. in *The Boxwood Bulletin,* Vol. 15(3):41-42, Jan. 1976.

History: Discovered as a mutation of *Buxus microphylla* 'Compacta' by Henry Hohman of Kingsville Nurseries, Kingsville, Md., in the early 1960s. Was named by Dr. J.T. Baldwin, Jr. for Mrs. Edgar M. Whiting who was associated with The American Boxwood Society as editor of *The Boxwood Bulletin.*

Additional Information:
Culture and Care: Prefers dappled shade; will tolerate siting in some direct sun but occasionally suffers from winter bronzing.
Pests and Diseases: Resistance to leaf miner, psyllid and mites.
Not available in the commercial nursery trade.

Buxus microphylla 'Henry Hohman'

Size (25 yrs.):	Medium – 4½ to 5 feet high; 5 to 6 feet wide.
Natural Form:	Unusual – loose, with distorted irregular branchlets.
Annual Growth:	Medium – 2 to 2½ inches high and 2½ to 2¾ inches wide.
Leaf Color:	Light yellow-green.
Leaf Shape:	Obovate, moderately revolute; obtuse tip, some retuse; cuneate base.
Leaf Size:	Small – ⅜ to ⅝ inch long and ¼ to 5/16 inch wide.
Leaf Surface:	Glabrous and smooth.
Internodal Length:	Short – ¼ to ⅜ inch.
Flowering Habit:	Sparse flowering and fruiting.
Hardiness:	Zones 6 to 9.
Plant Use:	Unusual specimen, espalier.

Registration: Not registered.
History: Discovered in the 1940s as a mutation of *Buxus microphylla* 'Compacta' (erroneously referred to as 'Kingsville Dwarf') by Henry Hohman of Kingsville Nurseries, Kingsville, Md. The loss of his usually assigned numerical designations while being grown-on and tested until named probably accounts for arboreta and collectors just calling it 'Henry Hohman.'
Bibliography: Not documented
Known Locations: College of William and Mary, U.S. National Arboretum, State Arboretum of Virginia.
Additional Information:
Culture and Care: Prefers dappled shade but will tolerate siting in direct sun; usually suffers from winter bronzing.
Pests and Diseases: Resistance to leaf miner, psyllid and mites.
Propagation: Difficult and quite slow to root. Take cuttings when new growth has hardened and include a portion of the heel from the previous year's growth; bruise the stem tissue; use an IBA powder dip.
Available in the commercial nursery trade in Europe.

Buxus microphylla 'Jim's True Spreader'

Size (25 yrs.):	Insufficient data to yet determine.
Natural Form:	Insufficient data. Indications are that it will most likely be in the mounded category but quite loose and of open habit.
Annual Growth:	Not documented
Leaf Color:	Medium green.
Leaf Shape:	Obovate tending toward rotund; obtuse tip and lightly retuse; cuneate base.
Leaf Size:	Medium – 3/4 to 1 1/8 inches long and 3/8 to 9/16 inch wide.
Leaf Surface:	Glossy, shiny and smooth.
Internodal Length:	Medium – 5/8 to 3/4 inch.
Flowering Habit:	Not observed.
Hardiness:	Believed to be USDA Zones 6 to 8.
Plant Use:	Specimen.

Registration: Not registered.
History: Released by Saunders Orchard & Nurseries, Piney River, Va. Originally from Garrison's Nursery, Seabreeze, N.J.
Bibliography: Not documented
Known Locations: State Arboretum of Virginia.
Additional Information:
Culture and Care: Demonstrates no special cultural requirements.
Pests and Diseases: Indicates resistance to leaf miner and mites in the more humid climates; no serious diseases.
Propagation: Cuttings root quite readily without any special preparations. The poly-tent procedure usually produces rooted cuttings in 7 to 8 weeks.
Available in the commercial nursery trade.

Buxus microphylla 'John Baldwin'

Size (25 yrs.):	Medium – 4½ to 5 feet high and 3½ to 4 feet wide. The parent clone, at 35 years, measures 10 feet high and 4½ feet wide.
Natural Form:	Pyramidal, billowy; somewhat prone to exhibit candle-like shoots. As the plant ages it forms a more conical shape.
Annual Growth:	Medium – 2 to 2½ inches high and 1½ to 2 inches wide.
Leaf Color:	Dark green, undertones of yellow, somewhat bluish cast, prone to bronze when sited in direct sunlight.
Leaf Shape:	Ovobate to lanceolate; obtuse tip, some retuse; cuneate base.
Leaf Size:	Small – ⅝ to ¾ inch long and ¼ to ⁵⁄₁₆ inch wide.
Leaf Surface:	Glabrous and smooth.
Internodal Length:	Medium – ¼ to ⅜ inch.
Flowering Habit:	Moderate flowering and fruiting.
Hardiness:	Zones 6 to 8.
Plant Use:	Specimen, grouping for background and area separations, hedgings.

Registration: P.D. Larson in *The Boxwood Bulletin*, Vol. 28(2):27, Oct. 1988.
History: Discovered in 1950 as an open-pollinated seedling in Colonial Williamsburg by Dr. John T. Baldwin, Jr. of the College of William and Mary. It was named by Dr. Bernice Speese of the College of William and Mary.
Bibliography:
The Boxwood Society of the Midwest Bulletin, April 1986.
The Boxwood Bulletin, Vol. 22(2):31, Oct. 1982/27(3): 65, Jan. 1988.
Known Locations: Buzzards Bay Garden Club, College of William and Mary, U.S. National Arboretum, State Arboretum of Virginia.
Additional Information:
Culture and Care: Tolerates direct sun, prefers some shade. I really like this plant.
    Pruning of the candle-like shoots improves the plant's appearance.
Pests and Diseases: Resistance to leaf miner, psyllid and mites.

Buxus microphylla 'Kingsville'

Size (25 yrs.):	Medium – 3 to 3½ feet high and 2½ to 3 feet wide.
Natural Form:	Mounded – loose habit; upright.
Annual Growth:	Slow – 1 to 1½ inches in height and ¾ to 1 inch in width.
Leaf Color:	Medium yellow-green.
Leaf Shape:	Lanceolate; ovate tip; cuneate base.
Leaf Size:	Large – 1 to 1⅛ inches long and ⅜ to ½ inch wide.
Leaf Surface:	Glossy and smooth.
Internodal Length:	Short – 3/16 to ¼ inch.
Flowering Habit:	Sparse flowering and not observed to have set fruit. This may change as the plant ages.
Hardiness:	Zones 6 to 8.
Plant Use:	Specimen, foundation planting.

4 ft

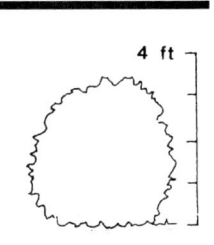

Registration: Not registered.
History: Believed to have originated with Henry Hohman in the 1940s at Kingsville Nurseries, Kingsville, Md. The loss of his usually assigned numerical designations until named probably accounts for growers just calling it 'Kingsville.'
Bibliography:
Everett, T.H. *The New York Botanical Garden Ill. Encyc. of Hort.*, Vol. 2, 1981.
Known Locations: U.S. National Arboretum, State Arboretum of Virginia.
Additional Information:
Culture and Care: Demonstrates no special cultural requirements. Requires full shade.
Pests and Diseases: Indicates resistance to leaf miner.
Not available in the commercial nursery trade.

Buxus microphylla 'Locket'

Size (25 yrs.):	Medium – 5 to 5½ feet tall and 3 to 3½ feet wide.
Natural Form:	Unusual – loose habit and somewhat twisted.
Annual Growth:	Medium – 2½ to 3 inches in height and 1 to 1¼ inches in width.
Leaf Color:	Light green.
Leaf Shape:	Obovate; obtuse tip with some retuse; cuneate base.
Leaf Size:	Small – ¼ to ½ inch long and ⅛ to 7/16 inch wide.
Leaf Surface:	Glabrous and smooth.
Internodal Length:	Short – ¼ to 5/16 inch.
Flowering Habit:	Moderate flowering and moderate fruiting.
Hardiness:	Zones 5(protected) to 8.
Plant Use:	Specimen.

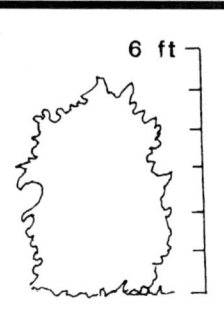

Registration: J.T. Baldwin in *The Boxwood Bulletin*, Vol. 15(3):41, 1976 and Vol. 16(1):10-11, 1976.

History: *Buxus microphylla* 'Locket' originated with Dr. J.T. Baldwin, Jr. of the College of William and Mary in the early 1950s from seed of *Buxus microphylla* 'Curly Locks.' Dr. Baldwin described 'Locket' as a most distinctive and valuable plant and vastly different from any other box that he knew. Believed to have been named for Miss Lucy Locket of poetical and musical fame.

Bibliography:
The Boxwood Bulletin, Vol. 15(3):41, Jan. 1976/16(1):10-11, July 1976/18(3):54, Jan. 1979/19(2):26, Oct. 1979.

Known Locations: College of William and Mary near the lower end of the Sunken Garden, Williamsburg, Va.

Additional Information:
Culture and Care: Demonstrates no special cultural requirements.
Pests and Diseases: Indicates resistance to leaf miner, psyllid and mites in the more humid climates; no serious diseases.
Not available in the commercial nursery trade.

Buxus microphylla 'Miss Jones'

Size (25 yrs.):	Small – 3 to 3^1/$_2$ feet high and 3^1/$_2$ to 4 feet wide.
Natural Form:	Mounded and tending toward vase-shaped.
Annual Growth:	Slow – ^{3}/$_4$ to 1 inch in height and ^{3}/$_4$ to 1 inch in width.
Leaf Color:	Medium yellow-green.
Leaf Shape:	Lanceolate; ovate tip; cuneate base.
Leaf Size:	Small – ^{5}/$_8$ to ^{3}/$_4$ inch long and ^{3}/$_{16}$ to ^{1}/$_4$ inch wide.
Leaf Surface:	Glabrous and smooth.
Internodal Length:	Medium – ^{1}/$_4$ to ^{3}/$_{16}$ inch.
Flowering Habit:	Moderate flowering; not observed to have set fruit but this may change as the plant ages.
Hardiness:	Zones 6 to 8.
Plant Use:	Specimen, grouping for background and area separations, edgings, hedgings.

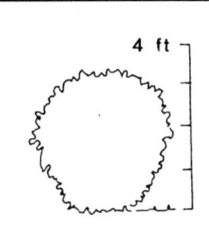

Registration: Not registered.
History: Believed to have originated with R. Jones, Route 2, Box 93, Eatonton, Ga., as an open-pollinated seedling and sent to Henry Hohman of Kingsville Nurseries, Kingsville, Md., where it was further propagated, named and commercially released.
Bibliography: Not documented
Known Locations: Washington Park Arboretum, U.S. National Arboretum, State Arboretum of Virginia.
Additional Information:
Culture and Care: Demonstrates no special cultural requirements.
Pests and Diseases: Indicates resistance to leaf miner, psyllid and mites in the more humid climates; no serious diseases.
Not yet available in the commercial nursery trade.

Selected Species, Varieties, Cultivars

Buxus microphylla 'Quiet End'

Size (25 yrs.):	Medium – 4 feet high; 8 feet wide.
Natural Form:	Mounded, somewhat loose and open.
Annual Growth:	Medium – 2 inches high; 4 inches wide.
Leaf Color:	Light yellow-green.
Leaf Shape:	Obovate; obtuse tip, some slightly retuse; cuneate base.
Leaf Size:	Small – $9/16$ to $11/16$ inch long; $5/16$ to $3/8$ inch wide.
Leaf Surface:	Glabrous – smooth.
Internodal Length:	Short – $1/4$ to $7/16$ inch.
Flowering Habit:	Not observed.
Hardiness:	Zones 6 to 8.
Plant Use:	Specimen; grouping for background and area separations; bonsai.

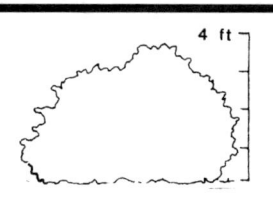

Registration: L.R. Batdorf in *The Boxwood Bulletin*, Vol. 31(3):32, Jan. 1992.
History: The plant originated as a mutation of *B. microphylla* 'Compacta.' It was selected by Henry Hohman at Kingsville Nursery in Kingsville, Md. Here the plant was originally known as *Buxus microphylla* 'Kingsville 2A.'
Bibliography:
The Boxwood Bulletin, Vol. 18(2):43, Oct. 1978.
Known Locations: U.S. National Arboretum, State Arboretum of Virginia.
Additional Information:
Culture and Care: Prefers dappled shade but will tolerate some direct sun. Will display bronze coloring during the winter months unless protected.
Pests and Diseases: Resistance to leaf miner, pysllid and mites.
Propagation: Difficult and quite slow to root. Take cuttings when spring growth has hardened and include a portion of the heel from the previous year's growth; bruise the stem tissue or slit each side of the stem; use an IBA powder dip.
Available in the commercial nursery trade as 'Kingsville 2A.'

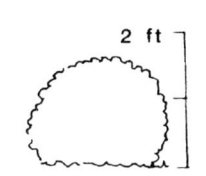

Buxus microphylla 'Sport Compacta No. 1'

Size (25 yrs.):	Dwarf – 18 to 20 inches high and 18 to 20 inches wide. A 42-year-old plant measures 3 feet high and 4 feet wide.
Natural Form:	Mounded, semi-globular.
Annual Growth:	Slow – $3/4$ to 1 inch high and $3/4$ to 1 inch wide.
Leaf Color:	Medium green.
Leaf Shape:	Lanceolate and slightly revolute; obtuse tip, some retuse; cuneate base.
Leaf Size:	Small – $1/2$ to $5/8$ inch long and $3/16$ to $5/16$ inch wide.
Leaf Surface:	Glabrous and smooth.
Internodal Length:	Short – $3/16$ to $1/4$ inch.
Flowering Habit:	Not observed.
Hardiness:	Zones 5 to 8.
Plant Use:	Specimen, edgings, bonsai.

Registration: Not registered.
History: Discovered and propagated by Dr. J.T. Baldwin, Jr. of the College of William and Mary in the late 1960s. Believed to be a mutation of *Buxus microphylla* 'Compacta,' it was propagated and released by Henry Hohman of Kingsville Nurseries, Kingsville, Md., as one of his numbered series.
Bibliography: Not documented
Known Locations: Colonial Williamsburg, State Arboretum of Virginia.
Additional Information:
Culture and Care: Prefers dappled shade but will tolerate siting in some direct sun.
Pests and Diseases: Resistance to leaf miner, psyllid and mites.
Not available in the commercial nursery trade.

Selected Species, Varieties, Cultivars

Buxus microphylla 'Sport Compacta No. 2'

Size (25 yrs.):	Dwarf – 18 to 20 inches high and 18 to 20 inches wide. A 42-year-old plant measures 3½ feet high and 4½ feet wide.
Natural Form:	Mounded tending toward globular.
Annual Growth:	Slow – ¾ to 1 inch high and ¾ to 1 inch wide.
Leaf Color:	Medium green.
Leaf Shape:	Slightly revolute; acute tip with some being obtuse; cuneate base.
Leaf Size:	Small – ½ to ¾ inch long and ³⁄₁₆ to ¼ inch wide.
Leaf Surface:	Glabrous and smooth.
Internodal Length:	Short – ³⁄₁₆ to ¼ inch.
Flowering Habit:	Not observed.
Hardiness:	Zones 5 to 8.
Plant Use:	Specimen, edgings, bonsai.

Registration: Not registered.
History: Discovered and propagated by Dr. J.T. Baldwin, Jr. of the College of William and Mary in the late 1960s. It is believed to be a mutation of *Buxus microphylla* 'Compacta.' A strong possibility exists that it was further propagated and released by Henry Hohman of Kingsville Nurseries, Kingsville, Md., as one of his numbered series.
Bibliography: Not documented
Known Locations: Colonial Williamsburg, State Arboretum of Virginia.
Additional Information:
Culture and Care: Demonstrates no special cultural requirements.
Pests and Diseases: Indicates resistance to leaf miner, psyllid and mites in the more humid climates; no serious diseases.
Not available in the commercial nursery trade.

Buxus microphylla 'Sunlight'

Size (25 yrs.):	Medium – 3½ to 4 feet high and 4½ to 5 feet wide.
Natural Form:	Mounded – semi-spherical, curves inward at base, somewhat loose and open habit.
Annual Growth:	Medium – 1½ to 1¾ inches high and 2 to 2½ inches wide.
Leaf Color:	Light yellow-green.
Leaf Shape:	Narrowly obovate with some being slightly revolute; obtuse tip with some being slightly retuse; cuneate base.
Leaf Size:	Small – ½ to 5/16 inch long and ¼ to 3/16 inch wide.
Leaf Surface:	Glossy and smooth.
Internodal Length:	Medium – 5/16 to 3/8 inch.
Flowering Habit:	Not observed.
Hardiness:	Zones 5 to 8.
Plant Use:	Specimen, grouping for background and area separations, edgings, hedgings.

Registration: Mary A. Gamble for The Boxwood Society of the Midwest in *The Boxwood Bulletin*, Vol. 28(2):26-27, Oct. 1988.
History: Believed to have emanated from a mutation of *Buxus microphylla* 'Compacta,' propagated and distributed by Dr. J.T. Baldwin, Jr. of the College of William and Mary in late 1969 as S-11.
Bibliography:
The Boxwood Bulletin, Vol. 28(2):26-29, Oct. 1988.
Known Locations: State Arboretum of Virginia.
Additional Information:
Culture and Care: Demonstrates no special cultural requirements.
Pests and Diseases: Indicates resistance to leaf miner, psyllid and mites in the more
    humid climates; no serious diseases.
Not available in the commercial nursery trade.

Buxus microphylla 'Sunnyside'

Size (25 yrs.):	Large – more than 6 feet high and 5½ feet wide.
Natural Form:	Pyramidal to mounded, loose open habit.
Annual Growth:	Fast – 3½ to 4 inches in height and 3 to 3½ inches in width.
Leaf Color:	Medium yellow-green.
Leaf Shape:	Rotund; obtuse tip with some being retuse; cuneate base.
Leaf Size:	Large – ⅞ to 1 inch long and ½ to ⅝ inch wide.
Leaf Surface:	Glossy and smooth.
Internodal Length:	Medium – ⅜ to ½ inch.
Flowering Habit:	Moderate flowering and moderate fruiting.
Hardiness:	Zones 5 to 8.
Plant Use:	Specimen, foundation planting, grouping for background and area separations.

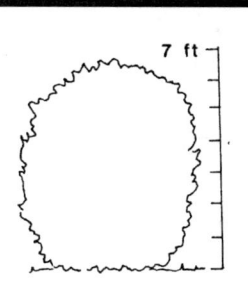

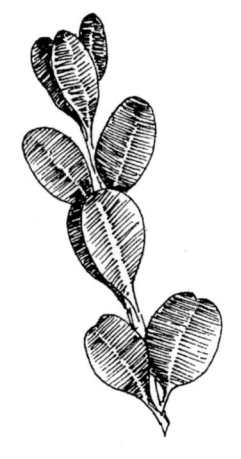

Registration: Not registered.
History: Named and commercially released by Sunnyside Nurseries, Troy, Ill.; also carried by John Vermeullen & Son, Neshanic Station, N.J., in 1984.
Bibliography:
Dirr, M.A. *Manual of Woody Landscape Plants*, 4th Ed., 1990.
Flint, H.L. *Horticulture,* March 1987.
Krussman, G. *Manual of Cultivated Broad-leaved Trees and Shrubs*, Vol. I, A-D, 1984.
Known Locations: Arnold Arboretum, U.S. National Arboretum, State Arboretum of Virginia.
Additional Information:
Culture and Care: Demonstrates no special cultural requirements.
Pests and Diseases: Indicates resistance to leaf miner, psyllid and mites in the more humid climates; no serious diseases.
Available in the commercial nursery trade of North America and Europe.

Buxus microphylla 'Winter Gem'

Size (25 yrs.):	Medium – 5 to 5^1/$_2$ feet high and 4 to 4^1/$_2$ feet wide.
Natural Form:	Pyramidal, loose and open habit.
Annual Growth:	Medium – 2^1/$_2$ to 3 inches high and 2 to 2^1/$_2$ inches wide.
Leaf Color:	Medium yellow-green.
Leaf Shape:	Ovate; obtuse tip, some retuse; cuneate base.
Leaf Size:	Medium – ^{7}/$_8$ to 1 inch long and ^{3}/$_8$ to ^{1}/$_2$ inch wide.
Leaf Surface:	Glossy and smooth.
Internodal Length:	Medium – ^{11}/$_{16}$ to ^{13}/$_{16}$ inch.
Flowering Habit:	Not observed.
Hardiness:	Zones 5 to 8.
Plant Use:	Specimen, foundation planting, grouping for background and area separations.

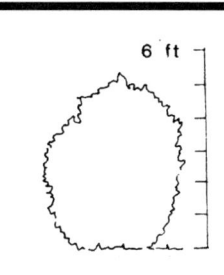

Registration: Not registered.
History: Commercially released by John Vermeullen & Son, Neshanic Station, N.J., 1982.
Bibliography: Not documented
Known Locations: Missouri Botanical Garden, U.S. National Arboretum, State Arboretum of Virginia.
Additional Information:
Culture and Care: Prefers dappled shade but will tolerate being sited in some direct sun. Breaks dormancy early.
Pests and Diseases: Resistance to leaf miner, psyllid and mites.

Buxus microphylla – Partial Descriptions

Additional cultivars with insufficient data for a plant description:

'Fiorii' – Catalog listing by Fiore Enterprises, Route 2, Prairie View, Ill.

'Green Cushion' – Inventory Coker Arboretum, University of North Carolina, Chapel Hill, N.C. (1988).

'Green Sofa' – Plant originated by Dr. John T. Baldwin, Jr. of the College of William and Mary from a mutation of *Buxus microphylla* 'Green Pillow,' *The Boxwood Bulletin*, Vol. 15(3):42, Jan. 1976. I here describe one of the sports as 'Green Sofa' – more vigorous than parent (*B. mic.* 'Green Pillow') and its leaves coarser; leaves deep green, elliptic (³/₄ inch long by ¹/₂ inch broad), bases acute, tips cuspidate. 'Green Pillow' does not flower; 'Green Sofa' is too young to assess in this respect.

'Kingsville HNS' – Inventory Secret Arboretum (1982).

'Kingsville L' – Inventory U.S. National Arboretum (4201).

'Kingsville 2' – Inventory U.S. National Arboretum.

'Kingsville 4' – Inventory Washington Park Arboretum (1983).

'Kingsville 4A' – Inventory U.S. National Arboretum (29692).

'Kingsville 58' – Inventory Secret Arboretum (1982).

'Morris Medium Dwarf' – Inventory Secret Arboretum (1982). *The Boxwood Bulletin*, Vol. 21(4):63, April 1982. "Grown for 12 years. Outplanted in dwarf evergreen garden as a five-inch container plant. A quarter of plant winter killed by temperatures to -20º F. Additional winter kill during three successive severe winters with temperatures from -17º F. to -20º F. although completely under snow cover. One third of plant killed. Recovers during spring growing season. Plant is presently (Nov. 1981) 9 inches high with a 14-inch spread. It is a dense compact mound. Tips of twigs up to 2 inches long have been killed by early October 1981 freezes with temperatures at 24º F. Plant is presently in poor condition. Marginally hardy only on protected sites."

'Morrison Garden' – Inventory Secret Arboretum (1982). *The Boxwood Bulletin*, Vol. 21(4):63, April 1982. "Grown for 12 years on a very well-protected site with light high shade. Plant holds green color through winter. No apparent damage with temperatures to -20º F. Plant is upright 39 inches tall with 28-inch spread. Plant is in excellent condition. Hardy on protected sites. Has increased its height by 4 inches a year."

'Ries Selection' – Inventory Secret Arboretum (1982). *The Boxwood Bulletin*, Vol. 21(4):63, April 1982. "This plant is a good example of effect of site. It grew for 30 years in a foundation planting on the east side of a house located in a deep ravine where it was completely protected from wind. Direct sunshine only reached it for a couple of hours each morning. It went through -20º F. temperatures three different winters with no damage. The plant was a miniature, being 10 inches tall with an 8-inch spread. In 1980 it was outplanted in a dwarf evergreen garden where it was exposed to some wind and winter sunshine. Eighty percent of the plant was killed the first winter with the lowest temperature being -14º F. Not hardy except on exceptionally protected sites moist but well-drained."

'True Spreader' – Catalog listing by Hines Nursery, Santa Fe, Calif., 1984. Could very well be 'Jim's True Spreader.'

'Wiertz' – Catalog, Firma C. Esvold, Boskoop, Holland, 1987.

Buxus microphylla variety *japonica* 'Green Beauty'

Size (25 yrs.):	Medium – 3½ to 4 feet high and 4 feet wide.
Natural Form:	Mounded – open loose habit.
Annual Growth:	Medium – 3 to 3½ inches high and 3 to 3½ inches wide.
Leaf Color:	Dark green.
Leaf Shape:	Lanceolate, tending toward ovate; obtuse tip, some retuse; cuneate base.
Leaf Size:	Medium – ¾ to ⅞ inch long and ⁷/₁₆ to ½ inch wide.
Leaf Surface:	Glossy and smooth.
Internodal Length:	Medium – ¼ to ⅜ inch.
Flowering Habit:	Not observed.
Hardiness:	Zones 6 to 8.
Plant Use:	Specimen, foundation planting, grouping for background and area separations, hedgings.

Registration: Not registered.
History: Taxonomic confusion appears to exist between this plant and *Buxus sempervirens* 'Green Beauty' released by Eastern Shore Nurseries, Easton, Md., in 1964.
Bibliography:
Dirr, M.A. *Manual of Woody Landscape Plants*, 4th Ed., 1990.
Gamble, M.A. *Flower and Garden,* March 1988.
Known Locations: State Arboretum of Virginia.
Additional Information:
Culture and Care: Prefers dappled shade; will tolerate being sited in some direct sun. Breaks dormancy early.
Pests and Diseases: Resistance to leaf miner, psyllid and mites.
Available in the commercial nursery trade.

Buxus microphylla variety *japonica* 'Morris Dwarf'

Size (25 yrs.):	Small – 2 to 2½ feet high and 2½ to 3 feet wide. A 42-year-old specimen measured 2¾ feet high, 5½ feet wide.
Natural Form:	Mounded – compact, dense and brittle.
Annual Growth:	Slow – 1 to 1¼ inches high and 1¼ to 1½ inches wide.
Leaf Color:	Medium green with yellow undertones.
Leaf Shape:	Obovate; obtuse tip, some retuse; cuneate base.
Leaf Size:	Small – ⁵⁄₁₆ to ½ inch long and ¼ to ⅜ inch wide.
Leaf Surface:	Glabrous and smooth.
Internodal Length:	Short – ³⁄₁₆ to ¼ inch.
Flowering Habit:	Not observed.
Hardiness:	Zones 5 to 8.
Plant Use:	Specimen, edgings, hedgings, bonsai.

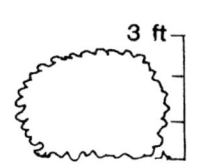

Registration: B. Wagenknecht in *The Boxwood Bulletin*, Vol. 11(3):45, Jan. 1972.
History: Discovered at The Morris Arboretum, Philadelphia, Pa., as a seedling of *Buxus microphylla* variety *japonica*. Selected by Dr. Henry T. Skinner in 1950 and named by the U.S. National Arboretum. At one time was named 'Morris Medium Dwarf.'
Bibliography:
The Boxwood Bulletin, Vol. 11(3):45, Jan. 1972/16(1):8,9, July 1976/17(2):25, Oct. 1977.
Known Locations: Arnold Arboretum, Longwood Gardens, Morris Arboretum, Washington Park Arboretum, U.S. National Arboretum, State Arboretum of Virginia.
Additional Information:
Culture and Care: Prefers dappled shade.
Pests and Diseases: Resistance to leaf miner, psyllid and mites.
Propagation: Difficult and quite slow to root. Take cuttings when spring growth has hardened and include a portion of the heel from the previous year's growth; bruise the stem tissue or slit each side of the stem; use an IBA powder dip.
Available in the commercial nursery trade.

Buxus microphylla variety *japonica* 'Morris Midget'

Size (25 yrs.):	Small – 2 to 2½ feet high and 2½ to 3 feet wide.
Natural Form:	Mounded and compact.
Annual Growth:	Slow – 1 to 1¼ inches high and 1 to 1½ inches wide.
Leaf Color:	Medium green with yellow undertones. Bronzes in the winter when sited in direct sun.
Leaf Shape:	Obovate; obtuse tip with some being retuse; cuneate base.
Leaf Size:	Small – 5/16 to ½ inch long and ¼ to 3/8 inch wide.
Leaf Surface:	Glabrous and smooth.
Internodal Length:	Short – 3/16 to ¼ inch.
Flowering Habit:	Not observed.
Hardiness:	Zones 5 to 8.
Plant Use:	Specimen, edgings, hedgings.

Registration: B. Wagenknecht in *The Boxwood Bulletin*, Vol. 11(3):45, Jan. 1972.
History: Originated at the Morris Arboretum, Philadelphia, Pa., from a seedling of *Buxus microphylla* variety *japonica*. Selected by Dr. Henry Skinner in 1950 and named by the U.S. National Arboretum.
Bibliography:
Dirr, M.A. *Manual of Woody Landscape Plants*, 4th Ed., 1990.
Flint, H.L. *Horticulture*, March 1987.
Missouri Botanical Garden Bulletin, LXXVI(3):8-10, May-June 1988.
The Boxwood Society of the Midwest Bulletin, April 1986.
The Boxwood Bulletin, Vol. 11(3):45, Jan. 1972/16(1):9, July 1976/17(2):25, Oct. 1977/ 20(3):43,46, Jan. 1981/27(3):65, Jan. 1988.
Known Locations: Arnold Arboretum, Longwood Gardens, Missouri Botanical Garden, Morris Arboretum, Washington Park Arboretum, U.S. National Arboretum, State Arboretum of Virginia.
Additional Information:
Culture and Care: Transplants readily; prefers dappled shade but will tolerate being sited in some direct sun but occasionally suffers from winter bronzing and slightly more yellowing. The addition of organic compost as a soil amendment and one inch of mulch adds to the health of the plant. Water seldom and thoroughly; an inch of water every two weeks is sufficient for sites with well-drained soil. Tolerates a pH range of slightly acidic to slightly alkaline with a preference for the sweet side (alkaline). Demonstrates no special cultural requirements.
Pests and Diseases: Indicates resistance to leaf miners, psyllid and mites in the more humid climates; no serious diseases.
Propagation: Somewhat difficult and quite slow to root. Best to take cuttings when spring growth has hardened and include a portion of the heel from the previous year's growth; bruise the stem tissue or slit each side of the stem; use an IBA powder dip.
Available in the commercial nursery trade.

Buxus microphylla variety *japonica* 'Nana Compacta'

Size (25 yrs.):	Dwarf – 10 to 12 inches high and 25 to 28 inches wide. The smallest boxwood that I know of.
Natural Form:	Mounded, twiggy and uneven.
Annual Growth:	Slow – less than $1/2$ inch high and $1/2$ to $3/4$ inch wide.
Leaf Color:	Medium green.
Leaf Shape:	Ovate; obtuse tip; cuneate base.
Leaf Size:	Small – $5/16$ to $3/8$ inch long and $3/16$ to $1/4$ inch wide.
Leaf Surface:	Glossy and smooth.
Internodal Length:	Short – $1/8$ to $3/16$ inch.
Flowering Habit:	Not observed.
Hardiness:	Zones 5(protected) to 8.
Plant Use:	Specimen, edgings, bonsai.

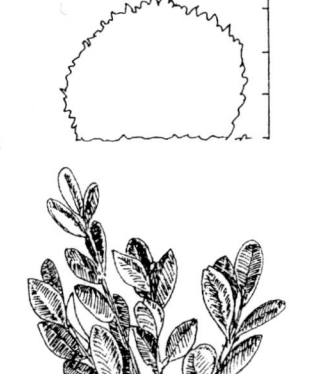

Registration: Catalog, Mayfair Nurseries, Bergenfield, N.J., 1954. Catalog, Mayfair Nurseries, Windham, Pa., 1972.
History: Discovered as a mutation of *Buxus microphylla* 'Compacta' some time in the 1940s. Named and commercially released by Walter A. Kolaga of Mayfair Nurseries, Bergenfield, N.J., in 1954.
Bibliography:
Wyman D. *Wyman's Gardening Encyclopedia*, New Expanded 2nd Ed. 1986.
Known Locations: Arnold Arboretum, State Arboretum of Virginia.
Additional Information:
Culture and Care: Prefers dappled shade.
Pests and Diseases: Resistance to leaf miner, psyllid and mites.
Propagation: Difficult and quite slow to root. Take cuttings when spring growth has hardened and include a portion of the heel from the previous year's growth; bruise the stem tissue or slit each side of the stem; use an IBA powder dip.
Available in the commercial nursery trade.

Buxus microphylla variety *japonica* 'National'

Size (25 yrs.):	Large – 9 to 10 feet high and 7 to 8 feet wide. A 36-year-old specimen measured 15¼ feet high and 15 feet wide.
Natural Form:	Pyramidal – somewhat open with straight ascending branches.
Annual Growth:	Fast – 4 to 5 inches in height and 3 to 4 inches in width.
Leaf Color:	Dark green; underside salmon color in the winter.
Leaf Shape:	Rotund and slightly revolute; retuse tip; cuneate base.
Leaf Size:	Large – 1 to 1³⁄₁₆ inches long and ½ to ⅝ inch wide.
Leaf Surface:	Glossy and smooth.
Internodal Length:	Long – ⁷⁄₁₆ to ¾ inch.
Flowering Habit:	Sparse flowering and sparse fruiting.
Hardiness:	Zones 6 to 8.
Plant Use:	Specimen, grouping for background and area separations, hedgings.

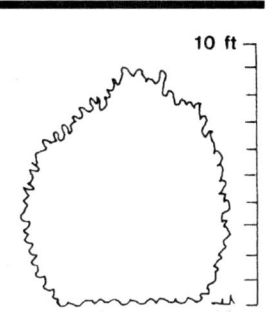

Registration: D. Anberg in *The Boxwood Bulletin,* 12(4):62, April 1973.
History: Originated as a seedling of *Buxus microphylla* variety *japonica* at Morris Arboretum, Philadelphia, Pa. Selected by Dr. Henry T. Skinner in 1951 and named by the U.S. National Arboretum. Once called 'Morris Fastigiate.'
Bibliography:
The Boxwood Bulletin, Vol. 12(4):62, April 1973/16(1):8-9, July 1976/17(2):25, Oct. 1977.
Known Locations: College of William and Mary, U.S. National Arboretum, State Arboretum of Virginia.
Additional Information:
Culture and Care: Demonstrates no special cultural requirements.
Pests and Diseases: Indicates resistance to leaf miner, psyllid and mites in the more humid climates; no serious diseases.
Available in the commercial nursery trade.

Buxus microphylla variety *japonica* – Partial Descriptions

Additional cultivars with insufficient data for a plant description:

'Alba' – Catalog, Andorra Nurseries, Chestnut Hill, Philadephia, Pa., 1908.

'Angustifolia' – L.H. Bailey in *Hortus Third*, 105, 1930. "with long narrow leaves."

'Argentea' – Beissner, Schelle and Zable in *Handbuch der Laubholz-Benennung*, 283, 1903.

'Aurea' – Catalog, Charles Dietriche, Angers, France, 1892.

'Fortunei' – Catalog, Andorra Nurseries, Chestnut Hill, Philadelphia, Pa., 1908.

'Japanese Globe' – Plant list, Kelly Howell, Spokane, Wash., 1958. "Faster growing than true dwarf. A border plant."

'Latifolia' – Catalog, Andorra Nurseries, Chestnut Hill, Philadelphia, Pa., 1908.

'Nana' – Beissner, Schelle and Zable in *Handbuch der Laubholz-Benennung*, 283, 1903.

'Obcordata' – Beissner, Schelle and Zable in *Handbuch der Laubholz-Benennung*, 283, 1903.

'Obcordata Variegata' – Anon. in List of Plants Introduced by Robert Fortune from Japan. *Gardeners Chronicle* 735, 1861. "A very pretty little variegated box with remarkably short, obtuse, sometimes retuse or obcordate leaves, of about half an inch in diameter."

'Rotundifolia' – Beissner, Schelle and Zable in *Handbuch der Laubholz-Benennung*, 283, 1903.

'Rotundifolia Glauca' – Catalog, Charles Dietriche, Angers, France, 1892.

'Rotundifolia Pendula' – Catalog, Andorra Nurseries, Chestnut Hill, Philadelphia, Pa., 1919.

'Rubra' – T. Makino in *Botanical Magazine of Tokyo*, 27:112, 1913. G. Krussman in *Manual of Cultivated Broad-leaved Trees and Shrubs*. "Garden form with orange-yellow leaves." "Small shrub, leaves oval to oblong, coriaceous, 3-12 mm long, 2-7-1/2 mm wide, orange colored, veins close. Flower small."

'Variegata' – *Hortus Third*, "leaves white- or yellowish-variegated."

Buxus sempervirens 'Abilene'

Size (25 yrs.):	Large – 7 to 8 feet high and 8 to 8½ feet wide.
Natural Form:	Pyramidal, billowy, tending toward ovate.
Annual Growth:	Fast – 3 to 4 inches high and 4 to 4½ inches wide.
Leaf Color:	Medium green with olive cast.
Leaf Shape:	Lanceolate, slightly revolute, some elliptic; acute tip, some obtuse; cuneate base.
Leaf Size:	Large – ⁷/₈ to 1⅛ inches long and ¼ to ½ inch wide.
Leaf Surface:	Glabrous and smooth.
Internodal Length:	Long – ½ to ⅝ inch.
Flowering Habit:	Not observed.
Hardiness:	Zones 5 to 8.
Plant Use:	Specimen, grouping for background and area separations, hedgings.

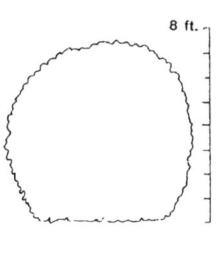

Registration: Inventory, Beal-Garfield Garden, East Lansing, Mich., 1960.
History: The original clone is believed to have come to Abilene, Kan., in about 1891 from a Pennsylvania nursery. Mrs. Emma Wolf purchased six of these boxwoods as a gift for her mother, Mrs. Katrina Hasbaggen, a native of Germany; only one plant survived to become the parent of the 'Abilene' clone.
Bibliography:
The Boxwood Bulletin, Vol. 11(3):38-40, Jan. 1972.
Known Locations: Grounds of Trinity Lutheran Church, Abilene, Kan.; Eisenhower Library; U.S. National Arboretum; State Arboretum of Virginia.
Additional Information:
Culture and Care: Tolerates siting in direct sun. Breaks dormancy late.
Pests and Diseases: Prone to leaf miner, psyllid and mites, but controllable.
Available in the commercial nursery trade.

Buxus sempervirens 'Agram'

Size (25 yrs.):	Medium – 5 to 6 feet high and 6 to 7 feet wide.
Natural Form:	Pyramidal, billowy.
Annual Growth:	Medium – 2 1/2 to 3 inches high and 3 to 3 1/2 inches wide.
Leaf Color:	Medium green with undertones of yellow.
Leaf Shape:	Lanceolate, some elliptic; acute tip; cuneate base.
Leaf Size:	Medium – 3/4 to 1 1/16 inches long and 5/16 to 3/8 inch wide.
Leaf Surface:	Glabrous, smooth.
Internodal Length:	Medium – 5/16 to 3/8 inch.
Flowering Habit:	Sparse flowering and fruiting.
Hardiness:	Zones 5 to 8.
Plant Use:	Specimen, grouping for background and area separations, hedgings.

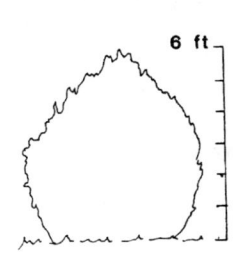

Registration: Introduced by the U.S. Department of Agriculture in 1959.
History: Seed was sent to Dr. Edgar Anderson from Skoplje, Vardar River Valley, Macedonia, Yugoslavia in 1935. Propagated at the Missouri Botanical Garden and designated within what was called the K-series. K-79 was considered outstanding and was named 'Agram' after the old geographic name for Zagreb.
Bibliography:
Krussman, G. *Manual of Cultivated Broad-leaved Trees and Shrubs*. Vol. I, A-D, 1984.
The Boxwood Society of the Midwest Bulletin, April 1986.
The Boxwood Bulletin, Vol. 8(4):63, April 1969/10(2):26-27, Oct. 1970/14(4):59-62, April 1975/24(2):50, Oct. 1984.
Known Locations: Arnold Arboretum, Buzzards Bay Garden Club, Hillier's Arboretum, Missouri Botanical Garden, Washington Park Arboretum, U.S. National Arboretum, State Arboretum of Virginia.
Additional Information:
Culture and Care: I really like this plant and it won't overwhelm smaller properties.
   Tolerates siting in direct sun quite well. Breaks dormancy late.
Pests and Diseases: Resistance to leaf miner, psyllid and mites.
Available in the commercial nursery trade.

Buxus sempervirens 'Arborescens'

Size (25 yrs.):	Large – 11 to 12 feet high and 8 to 9 feet wide. A 54-year-old specimen measured 14 feet high and 10 feet wide. A 150-year-old specimen measured slightly more than 32 feet high.
Natural Form:	Arboreal with a tendency toward loosely pyramidal.
Annual Growth:	Fast – 5 to 6 inches in height and 4 to 5 inches in width.
Leaf Color:	Medium green.
Leaf Shape:	Ovate to ovate-elliptic and ovate-lanceolate; acute tips with some being slightly retuse; cuneate base.
Leaf Size:	Large – $3/4$ to 1 inch long and $1/4$ to $1/2$ inch wide.
Leaf Surface:	Glabrous and smooth.
Internodal Length:	Long – $3/8$ to $5/8$ inch.
Flowering Habit:	Sparse flowering and fruiting.
Hardiness:	Zones 5 to 8.
Plant Use:	Specimen, grouping for background and area separations, allees.

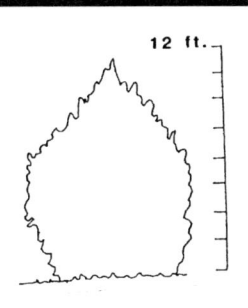

Registration: P. Miller in *Gardener's Dictionary*, ed. 8:Bux no. 1. 1756.
History: Most likely originated as an open-pollinated seedling somewhere in Europe.
Bibliography:
Bailey, L.H. *Hortus Third*, 1976.
Dallimore, W. *Holly, Yew and Box*, 1908.
Everett, T.H. *The New York Botanical Garden Ill. Encyc. of Hort.*, Vol. 2, 1981.
Flint, H.L. *Horticulture*, March 1987.
Hottes, A.C. *The Book of Shrubs*, 1931.
Krussman, G. *Manual of Cultivated Broad-leaved Trees and Shrubs*, Vol. I, A-D, 1984.
Wyman, D. *Wyman's Gardening Encyclopedia*, New Expanded 2nd Ed., 1986.
The Boxwood Bulletin, Vol. 2(3):36, Jan. 1963/3(2):24, Oct. 1963/7(2):17, Oct. 1967/ 8(4):62-63, April 1969/13(2):20-21, Oct. 1973/16(1):2, July 1976/17(2):25, Oct. 1977/20(2):31, Oct. 1980/20(4):80, April 1981/21(4):64, April 1982/22(4):70, April 1983/27(3):65, Jan. 1988.
Known Locations: Hillier's Arboretum, Royal Botanic Gardens, Kew, Longwood Gardens, Tennessee Botanical Gardens, Washington Park Arboretum, U.S. National Arboretum, State Arboretum of Virginia.

Additional Information:
Culture and Care: Demonstrates no special cultural requirements.
Pests and Diseases: Indicates an attraction for leaf miner and psyllid, with resistance to mites in the more humid climates; no serious diseases.
Available in the commercial nursery trade.

Buxus sempervirens 'Argenteo-variegata'

Size (25 yrs.):	Small – 2½ to 3 feet high and 4 to 5 feet wide.
Natural Form:	Mounded.
Annual Growth:	Slow – 1 to 1½ inches high and 2 to 2¼ inches wide.
Leaf Color:	Medium green with silver variegation most prominent at the leaf apex.
Leaf Shape:	Lanceolate, slightly revolute, some distorted; obtuse tip tending toward acute; cuneate base.
Leaf Size:	Large – 1 to 1¹/₁₆ inches long and ⅜ to ½ inch wide.
Leaf Surface:	Bullate, slightly puckered.
Internodal Length:	Medium – ¼ to ⅜ inch.
Flowering Habit:	Not observed.
Hardiness:	Zones 6 to 8.
Plant Use:	Specimen, grouping for background and area separations, hedgings.

Registration: R. Weston, *Botanicus Universalis* 1:31, 1770. Invalid names: 'Argenteo-marginata,' 'Suffruticosa Variegata Maculata,' 'Argentea.'
History: Most likely originated as an open-pollinated seedling or mutation somewhere in Europe.
Bibliography:
Bailey, L.H. *Hortus Third*, 1976.
Dallimore, W. *Holly, Yew and Box*, 1908.
Dirr, W.A. *Manual of Woody Landscape Plants*, 4th Ed., 1990.
Everett, T.H. *The New York Botanical Garden Ill. Encyc. of Hort.*, Vol. 2, 1981.
Hottes, A.C. *The Book of Shrubs*, 1931.
Krussman, G. *Manual of Cultivated Broad-leaved Trees and Shrubs*, Vol. I, A-D, 1984.
Wyman, D. *Wyman's Gardening Encyclopedia*, New Expanded 2nd Ed., 1986.
Known Locations: Brooklyn Botanic Garden, Buzzards Bay Garden Club, College of William and Mary, Hillier's Arboretum, Royal Botanic Gardens, Kew, Washington Park Arboretum, U.S. National Arboretum, State Arboretum of Virginia.
Additional Information:
Culture and Care: Prefers dappled shade, will tolerate siting in some direct sun, which emphasizes the variegation.
Pests and Diseases: Resistance to leaf miner, psyllid and mites.
Propagation: Difficult and quite slow to root. Take cuttings when new spring growth has hardened, taking the cuttings slightly below a node; bruise the stem or slit each side of the stem; use an IBA powder dip.
Available in the commercial nursery trade.

Buxus sempervirens 'Aristocrat'

Size (25 yrs.):	Large – 10 to 11 feet high; 6 to 7 feet wide. Two specimen plants at some 40 years old measure 14½ feet high and 13 feet wide.
Natural Form:	Pyramidal tending toward conical, dense habit.
Annual Growth:	Fast – 5 to 5½ inches in height and 3 to 3½ inches in width.
Leaf Color:	Medium green.
Leaf Shape:	Lanceolate; acute tip; cuneate base.
Leaf Size:	Medium – ¾ to ⅞ inch long and ⁵⁄₁₆ to ⁷⁄₁₆ inch wide.
Leaf Surface:	Glossy and smooth.
Internodal Length:	Long – ⁵⁄₁₆ to ⁷⁄₁₆ inch.
Flowering Habit:	Not observed.
Hardiness:	Zones 5(protected) to 8.
Plant Use:	Specimen, grouping for background and area separations, allees.

Registration: Dr. J.T. Baldwin, Jr. in *The Boxwood Bulletin,* Vol. 6(2):23, Oct. 1966.
History: The original two plants were purchased by Dr. J.T. Baldwin, Jr. of the College of William and Mary in 1953 from Justin B. Brouwers who had selected and grown them just outside Williamsburg, Va. They are still growing on the campus of the College of William and Mary, having been successfully moved from their original sites in early 1989.
Bibliography:
The Boxwood Society of the Midwest Bulletin, April 1986.
The Boxwood Bulletin, Vol. 6(2):23, Oct. 1966/14(1):10,15, July 1974.
Known Locations: College of William and Mary, Longwood Gardens, Secrest Arboretum, U.S. National Arboretum, State Arboretum of Virginia.
Additional Information:
Culture and Care: Tolerates siting in direct sun quite well. Breaks dormancy late.
Pests and Diseases: Resistance to leaf miner, psyllid and mites.
Not available in the commercial nursery trade.

Buxus sempervirens 'Asheville'

Size (25 yrs.):	Medium – 5½ to 6 feet high and 3½ to 4 feet wide.
Natural Form:	Pyramidal, somewhat billowy with a compact habit.
Annual Growth:	Medium – 2 to 2¼ inches in height and 1½ to 1¾ inches in width.
Leaf Color:	Dark green.
Leaf Shape:	Lanceolate; acute tip and slightly revolute; cuneate base.
Leaf Size:	Medium – ¾ to 13/16 inch long and ⅜ to ½ inch wide.
Leaf Surface:	Glabrous and smooth.
Internodal Length:	Medium – ½ to 13/16 inch.
Flowering Habit:	Not observed.
Hardiness:	Zones 5 to 8.
Plant Use:	Specimen, foundation planting, grouping for background and area separations, hedgings.

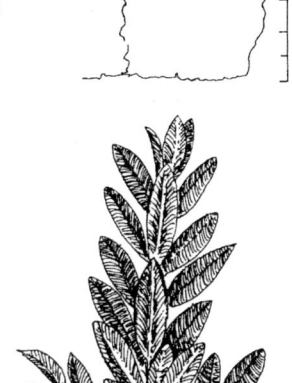

Registration: Not registered.
History: The original plant was growing on the grounds of the summer home formerly occupied by a son of President Hayes and located in West Asheville, N.C. Records indicate that 'Asheville' was being grown at Secrest Arboretum in Wooster, Ohio, in the early 1920s.
Bibliography:
The Boxwood Bulletin, Vol. 13(3):42-45, Jan. 1974/13(4):63, April 1974/16(1):14, July 1976/21(4):64,68, April 1982/22(3):49, Jan. 1983.
Known Locations: Secrest Arboretum.
Additional Information:
Culture and Care: Demonstrates no special cultural requirements.
Not available in the commercial nursery trade.

Buxus sempervirens 'Aurea Pendula'

Size (25 yrs.):	Large – 8 to 9 feet high and 10 to 11 feet wide.
Natural Form:	Unusual, branches are long and lax with the branchlets being pendulous.
Annual Growth:	Fast – 4 to 5 inches in height and 5 to 6 inches in width.
Leaf Color:	Bi-colored, gold-margined and dark green.
Leaf Shape:	Ovate to ovate elliptic and revolute; acute to obtuse tip; cuneate base.
Leaf Size:	Medium – $3/4$ to 1 inch long and $5/16$ to $3/8$ inch wide.
Leaf Surface:	Glabrous and smooth.
Internodal Length:	Long – $3/4$ to $7/8$ inch.
Flowering Habit:	Not observed.
Hardiness:	Zones 6 to 8.
Plant Use:	Specimen.

Registration: *Kew Handlist of Trees and Shrubs Grown in Arboretum*, Part II:131, 1896. Invalid 'Aurea Maculata Pendula.'
History: Probably originated as an open-pollinated seedling or mutation somewhere in Europe.
Bibliography:
Dallimore, W. *Holly, Yew and Box*, 1908.
Krussman, G. *Manual of Cultivated Broad-leaved Trees and Shrubs*, Vol. I, A-D, 1986.
The Boxwood Bulletin, Vol. 8(4):63, April 1969/17(2):25, Oct. 1977/20(4):80, April 1981.
Known Locations: Hillier's Arboretum, U.S. National Arboretum.
Additional Information:
Culture and Care: Demonstrates no special cultural requirements.
Pests and Diseases: Demonstrates resistance to leaf miner and psyllid; no serious diseases.
Available in the commercial nursery trade.

Buxus sempervirens 'Aureo-variegata'

Size (25 yrs.):	Large – 7 to 8 feet high and 7 to 8 feet wide.
Natural Form:	Pyramidal, billowy and open habit.
Annual Growth:	Fast – $4^1/_2$ to 5 inches in height and $4^1/_2$ to 5 inches in width.
Leaf Color:	Dark green with yellow speckles and variegations; young leaves usually yellow with faint variegations turning to green as they mature, particulary when growing in deep shade. As leaves age, much of the variegation is lost.
Leaf Shape:	Ovate to elliptic-oblong and somewhat revolute and sometimes distorted; acute to obtuse tip; cuneate base.
Leaf Size:	Large – 1 to $1^1/_8$ inches long and $^1/_2$ to $^5/_8$ inch wide.
Leaf Surface:	Glabrous and smooth.
Internodal Length:	Long – $^3/_8$ to $^5/_8$ inch.
Flowering Habit:	Sparse flowering and fruiting.
Hardiness:	Zones 6 to 8.
Plant Use:	Specimen, foundation planting, hedging, grouping for background and area separations.

Registration: R. Weston in *Botanicus Universalis*, 1:31, 1770. Excluded cultivar names: 'Arborescens Aurea Maculata,' H. Baillon, *Monographie des Buxacées et Stylocérées*, 60.1859; 'Aurea,' J. Loudon, *Arboretum et Fruticum Britannicum*, III:1333, 1838; 'Aurea Maculata,' *Kew Handlist of Trees and Shrubs Grown in Arboretum*, Part II:131, 1896; 'Aurea Maculata Aurea,' Inventory, Beal-Garfield Botanic Garden, East Lansing, Mich., 1960.
History: Probably originated somewhere in Europe as an open-pollinated seedling. Often seen in the older gardens.

Bibliography:
Bailey, L.H. *Hortus Third*, 1976.
Dallimore, W. *Holly, Yew and Box*, 1908.
Dirr, M.A. *Manual of Woody Landscape Plants*, 4th Ed., 1990.
Hottes, A.C. *The Book of Shrubs*, 1931.
Kew Handlist of Trees and Shrubs Grown in Arboretum, Part II:131, 1896.
Krussman, G. *Manual of Cultivated Broad-leaved Trees and Shrubs,* Vol. I, A-D, 1986.
Wyman, D. *Wyman's Gardening Encyclopedia,* New Expanded 2nd Ed., 1986.
The Boxwood Bulletin, Vol. 2(3):37, Jan. 1963/3(2):24, Oct. 1963/8(4):63, April 1969/ 17(2):25, Oct. 1977/20(4):80, April 1981/21(3):45, Jan. 1982.

Known Locations: Arnold Arboretum; Buzzards Bay Garden Club; Herrenhausen Garden, Hanover, Germany; Hillier's Arboretum; Royal Botanic Gardens, Kew; Longwood Gardens; Secrest Arboretum; U.S. National Arboretum; State Arboretum of Virginia.

Additional Information:
Culture and Care: Demonstrates no special cultural requirements.
Pests and Diseases: Indicates resistance to leaf miner, psyllid and mites in the more humid climates; no serious diseases.
Available in the commercial nursery trade.

Buxus sempervirens 'Belleville'

Size (25 yrs.):	Large – 7 to 8 feet high; 8 to 9 feet wide.
Natural Form:	Pyramidal – billowy, compact.
Annual Growth:	Medium – 2½ to 3 inches in height and 3½ to 4 inches in width.
Leaf Color:	Dark green, bluish cast to new growth.
Leaf Shape:	Elliptic; acute tip; cuneate base.
Leaf Size:	Medium – 7/16 to 3/4 inch long and 1/4 to 3/8 inch wide.
Leaf Surface:	Glabrous and smooth; tending toward dull.
Internodal Length:	Medium – 1/4 to 5/16 inch.
Flowering Habit:	Moderate flowering and fruiting.
Hardiness:	Zones 5 to 8.
Plant Use:	Specimen, grouping for background and area separations, hedgings.

Registration: R. Seibert in *Arnoldia*, Vol. 23(9):116, 1963.
History: The original clone was obtained by Mrs. Erwin W. Seibert in 1931 from Nick Bassler, a nurseryman near Belleville, Ill., who said the parent plant was growing near Scott Air Force Base on Route 2 near Belleville, Ill.
Bibliography:
Arnoldia, Oct. 1963.
Missouri Botanical Garden Bulletin, Vol. LXXVI(3):8-10, May-June 1988.
The Boxwood Society of the Midwest Bulletin, Supp. No. 1, April 1986.
The Boxwood Bulletin, Vol. 4(4):67, April 1965/20(3):43-45, Jan. 1981.
Known Locations: Buzzards Bay Garden Club, Missouri Botanical Garden, Washington Park Arboretum, U.S. National Arboretum, State Arboretum of Virginia.
Additional Information:
Culture and Care: An excellent plant but will overwhelm small properties unless kept pruned. Tolerates siting in direct sun quite well. Breaks dormancy late.
Pests and Diseases: Resistance to leaf miner, psyllid and mites.
Available in the commercial trade.

Buxus sempervirens 'Blauer Heinz'

Size (25 yrs.):	Dwarf – 2 to 2½ feet high and 2 to 2½ feet wide.
Natural Form:	Mounded, compact, dome-shaped habit, branches upright and erect.
Annual Growth:	Slow – ¾ to 1 inch in height and 1 to 1½ inches in width.
Leaf Color:	Dark green with bluish hue.
Leaf Shape:	Obovate; obtuse tip; cuneate base.
Leaf Size:	Small – ⅜ to ½ inch long and ³⁄₁₆ to ⁵⁄₁₆ inch wide.
Leaf Surface:	Glabrous and smooth.
Internodal Length:	Short – ³⁄₁₆ to ¼ inch.
Flowering Habit:	Not observed.
Hardiness:	Zones 5 to 8.
Plant Use:	Specimen, edgings, hedgings, bonsai.

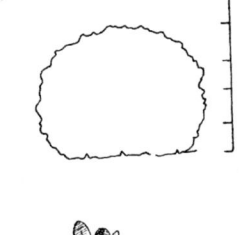

Registration: Otto Markworth and Dr. Hans-Georg Preissel in *Deutsche Baumschelle*, p. 516, Dec. 1987.
History: This clone was selected from a mixture of plants grown from cuttings originating in 1972 from several house-gardens in the Netherlands. It was named partly for its blue-green foliage and partly out of respect for its propagator, Heinz Grupe, who worked for Herrenhausen Garden in Hanover, Germany, for 40 years.
Bibliography:
The Boxwood Bulletin, Vol. 28(1):8-9, July 1988.
Known Locations: Herrenhausen Garden, U.S. National Arboretum, State Arboretum of Virginia.
Additional Information:
Culture and Care: Demonstrates no special cultural requirements.
Pests and Diseases: Exhibits resistance to leaf miner, no serious diseases.
Not available in the commercial nursery trade.

Buxus sempervirens 'Bullata'

Size (25 yrs.):	Large – 8 to 8½ feet high and 10 to 11 feet wide.
Natural Form:	Pyramidal with open and loose habit.
Annual Growth:	Fast – 3½ to 4 inches in height and 4½ to 5 inches in width.
Leaf Color:	Dark green.
Leaf Shape:	Lanceolate tending toward ovate and moderately revolute; acute tip; cuneate base.
Leaf Size:	Large – ⁷/₈ to 1¼ inches long and ³/₈ to ⁵/₈ inch wide.
Leaf Surface:	Bullate and slightly puckered.
Internodal Length:	Long – ³/₈ to ⁵/₈ inch.
Flowering Habit:	Floriferous and heavy fruiting.
Hardiness:	Zones 6 to 8.
Plant Use:	Specimen, grouping for background and area separations.

Registration: G. Kirchner in Petzold and Kirchner in *Arboretum Muscaviense*, 194, 1864.
History: Most likely originated as an open-pollinated seedling somewhere in Europe.
Bibliography:
Bailey, L.H. *Hortus Third*, 1976.
Dallimore, W. *Holly, Yew and Box*, 1908.
Dirr, M.A. *Manual of Woody Landscape Plants*, 4th Ed., 1990.
Everett, T.H. *The New York Botanical Garden Ill. Encyc. of Hort.*, Vol. 2, 1981.
Hottes, A.C. *The Book of Shrubs*, 1931.
Krussman, G. *Manual of Cultivated Broad-leaved Trees and Shrubs*, Vol. I, A-D, 1984.
Wyman, D. *Wyman's Gardening Encyclopedia*, New Expanded 2nd Ed., 1986.
The Boxwood Bulletin, Vol. 2(3):37, Jan. 1963/4(3):38, Jan. 1965/9(1):7, July 1969/ 17(2):25, Oct. 1977/21(3):45, Jan. 1982/27(3):65, Jan. 1988.
Known Locations: Arnold Arboretum, College of William and Mary, Royal Botanic Gardens, Edinburgh, Hillier Arboretum, Royal Botanic Gardens, Kew, Longwood Gardens, Washington Park Arboretum, U.S. National Arboretum, State Arboretum of Virginia.
Additional Information:
Culture and Care: Demonstrates no special cultural requirements.
Pests and Diseases: Indicates resistance to leaf miner, psyllid and mites in the more humid climates; no serious diseases.
Propagation: Cuttings root quite readily without the use of an IBA powder dip; however, they root slightly faster with the dip. The poly-tent procedure usually produces rooted cuttings nearly as fast as the mist systems, about 6 to 8 weeks.
Available in the commercial nursery trade.

Selected Species, Varieties, Cultivars

Buxus sempervirens 'Butterworth'

Size (25 yrs.):	Large – 7½ to 8 feet high and 6½ to 7 feet wide.
Natural Form:	Ovate with compact habit.
Annual Growth:	Medium – 3 to 3½ inches in height and 2½ to 3 inches in width.
Leaf Color:	Medium green with yellow undertone.
Leaf Shape:	Lanceolate; acute tips, some ovate and retuse; cuneate base.
Leaf Size:	Large – ⅝ to ⅞ inch long and ⅜ to ½ inch wide.
Leaf Surface:	Glossy and smooth.
Internodal Length:	Long – ½ to ⅝ inch.
Flowering Habit:	Sparse flowering; not observed to have set fruit.
Hardiness:	Zones 6 to 8.
Plant Use:	Specimen, grouping for background and area separations, hedgings.

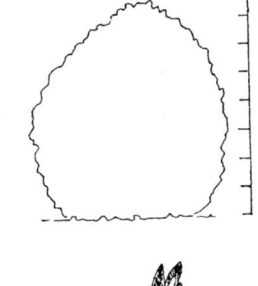

Registration: Catalog, Tingle Nurseries, Pittsfield, Md., 1958.
History: John Butterworth (1785-1840) lived at Butterworth Bridge, Petersburg, Va., and records indicate that he first grew the plant. His son continued to take cuttings from the original stands and lined them out in a field at his Osage Farm, Dinwiddie County, Va. Some years later, Tingle Nurseries of Pittsfield, Md., listed 'Butterworth' in its catalog; it is believed to have come circuitously from the original plants.
Bibliography:
The Boxwood Bulletin, Vol. 5(1):6-7, July 1965.
Known Locations: Arnold Arboretum, Buzzards Bay Garden Club, U.S. National Arboretum, State Arboretum of Virginia.
Additional Information:
Culture and Care: An excellent plant; will overwhelm small properties unless pruned. Tolerates being sited in some direct sun; occasionally suffers from winter bronzing. Breaks dormancy early.
Pests and Diseases: Resistance to leaf miner, psyllid and mites.
Available in the commercial nursery trade.

Buxus sempervirens 'Clembrook'

Size (25 yrs.):	Medium – 4 to 4½ feet tall and 3½ to 4 feet wide.
Natural Form:	Somewhat pyramidal, compact with upright branches.
Annual Growth:	Medium – 2 to 2½ inches in height and 1½ to 2 inches in width.
Leaf Color:	Dark green.
Leaf Shape:	Lanceolate and slightly revolute; obtuse tending toward acute; cuneate base.
Leaf Size:	Medium – ½ to ⅞ inch long and ¼ to ⅜ inch wide.
Leaf Surface:	Glabrous and smooth.
Internodal Length:	Medium – ¼ to ⅜ inch.
Flowering Habit:	Not observed.
Hardiness:	Zones 5 to 8.
Plant Use:	Specimen, foundation planting, grouping for background and area separations, hedgings.

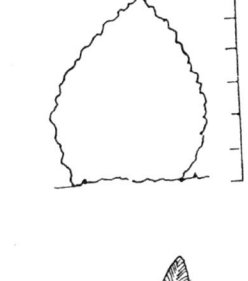

Registration: E. Bradford Clements in *The Boxwood Bulletin*, Vol. 19(2):26, Oct. 1979.
History: The plant was discovered about 1928 near Milton, Ontario, Canada, and many plants have been propagated from the original clone.
Bibliography:
The Boxwood Bulletin, Vol. 8(2):20-22, Oct. 1968/19(2):26, Oct. 1979.
Known Locations: State Arboretum of Virginia.
Additional Information:
Culture and Care: Demonstrates no special cultural requirements.
Pests and Diseases: Indicates resistance to leaf miner, psyllid and mites in the more humid climates; no serious diseases.
Available in the commercial nursery trade.

Buxus sempervirens 'Cliffside'

Size (25 yrs.):	Large – 9 to 10 feet high and 3 to 4 feet wide.
Natural Form:	Conical with stiff upright growth.
Annual Growth:	Fast – 4 to 4$^1/_2$ inches in height and 1 to 1$^1/_2$ inches in width.
Leaf Color:	Medium green.
Leaf Shape:	Elliptic and somewhat uneven; acute tip; cuneate base.
Leaf Size:	Medium – $^3/_4$ to $^{13}/_{16}$ inch long and $^5/_{16}$ to $^5/_8$ inch wide.
Leaf Surface:	Glossy and smooth.
Internodal Length:	Medium – $^1/_4$ to $^1/_2$ inch.
Flowering Habit:	Not observed.
Hardiness:	Zones 6 to 8.
Plant Use:	Specimen, grouping for background and area separations, hedgings, topiary.

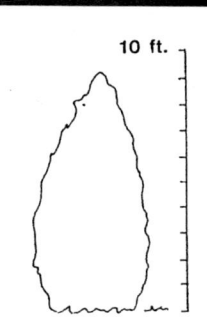

Registration: Dr. J.T. Baldwin, Jr. in *The Boxwood Bulletin*, Vol. 14(1):15, July 1974.
History: Dr. J.T. Baldwin, Jr. of the College of William and Mary discovered the open-pollinated seedling at a house site overlooking the Staunton River in Charlotte County, Va., in 1954.
Bibliography:
The Boxwood Bulletin, Vol. 14(1):14-16, July 1974/20(3): 42,45, Jan. 1981.
Known Locations: College of William and Mary, State Arboretum of Virginia.
Additional Information:
Culture and Care: Prefers dappled shade, will tolerate being sited in some direct sun, occasionally suffers from winter bronzing.
Pests and Diseases: Resistance to leaf miner, psyllid and mites.
Not available in the commercial nursery trade.

Buxus sempervirens 'Decussata'

Size (25 yrs.):	Large – 8 to 9 feet high and 5½ to 6 feet wide.
Natural Form:	Pyramidal with loose rather open habit.
Annual Growth:	Fast – 4 to 4½ inches in height and 2½ to 3 inches in width.
Leaf Color:	Dark green with a bluish hue.
Leaf Shape:	Lanceolate, slightly revolute, some slightly distorted; acute tip with some being obtuse; cuneate base.
Leaf Size:	Large – 1 to 1⅛ inches long and ½ to ¹¹⁄₁₆ inch wide.
Leaf Surface:	Glaucous, powdery coating.
Internodal Length:	Long – ⅜ to ⁷⁄₁₆ inch.
Flowering Habit:	Moderate flowering and moderate fruiting.
Hardiness:	Zones 6 to 8.
Plant Use:	Specimen.

Registration: L. Dippel in *Handbuch der Laubholzkunde*, 3:82, 1893.
History: Most likely originated as an open-pollinated seedling somewhere in Europe.
Bibliography: Not documented
Known Locations: Arnold Arboretum, Royal Botanic Gardens, Kew, Washington Park Arboretum, U.S. National Arboretum, State Arboretum of Virginia.
Additional Information:
Culture and Care: Demonstrates no special cultural requirements.
Pests and Diseases: Indicates resistance to leaf miner, psyllid and mites in the more humid climates; no serious diseases.
Not available in the commercial nursery trade.

Selected Species, Varieties, Cultivars

Buxus sempervirens 'Dee Runk'

Size (25 yrs.):	Large – 10 to 12 feet high and 2 to 2½ feet wide.
Natural Form:	Columnar – single vertical trunk having short branches growing upright.
Annual Growth:	Fast – 5½ to 6 inches in height and 1 to 1¼ inches in width.
Leaf Color:	Medium green.
Leaf Shape:	Elliptic; acute tip, some obtuse and slightly retuse; cuneate base.
Leaf Size:	Large – ⁷/₈ to 1³/₁₆ inches long and ³/₈ to ½ inch wide.
Leaf Surface:	Glabrous and smooth.
Internodal Length:	Medium – ½ and ¾ inch.
Flowering Habit:	Moderate flowering and fruiting.
Hardiness:	Zones 6 to 8.
Plant Use:	Specimen, grouping for background and area separations, allee.

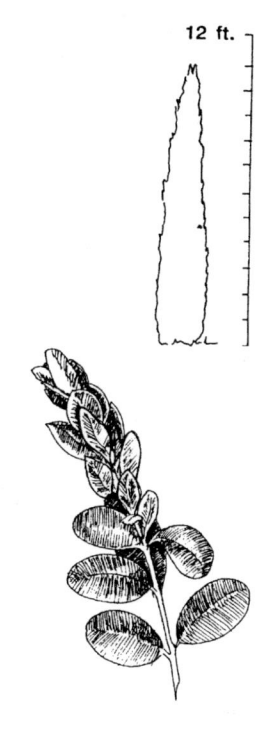

12 ft.

Registration: Mr. and Mrs. Charles K. Woltz in *The Boxwood Bulletin*, Vol. 28 (2):26, Oct. 1988.
History: The original plant (now lost) was located at Blandy Experimental Farm, home of the State Arboretum of Virginia, Boyce, Va. Cuttings were rooted in 1966 by Mr. and Mrs. Woltz of Charlottesville, Va. The cultivar was named for B.F.D. Runk, retired dean of students at the University of Virginia, former Samuel Miller Professor of Biology, and at one time associated with Blandy Experimental Farm.
Bibliography:
The Boxwood Bulletin, Vol. 27(3):52, Jan. 1988/28(2):26, Oct. 1988.
Known Locations: State Arboretum of Virginia.
Additional Information:
Culture and Care: Prefers dappled shade but will tolerate being sited in some direct sun, occasionally suffers from winter bronzing.
Pests and Diseases: Prone to leaf miner and psyllid.
Not available in the commercial nursery trade.

Buxus harlandii, weeping form, 3³/₄ ft. h x 5¹/₂ ft. w, age unknown, State Arboretum of Virginia. Photo by P. D. Larson

Buxus sinica variety *insularis* 'Justin Brouwers', 2¹/₂ ft. h x 3 ft. w, age 18 years, State Arboretum of Virginia. Photo by P. D. Larson

Buxus sinica variety *insularis* 'Tide Hill', 26 in. h x 78 in. w, age 45 to 50 years, Gloucester County, Virginia. Photo by P. D. Larson

Buxus ×
'Green Gem',
4½ h x 4¼ w,
age 20 years,
State Arboretum of
Virginia.
Photo by
C. A. Flangan

Buxus microphylla
variety *japonica*
'Morris Midget',
1½ ft. h x 2½ ft. w,
age 27 years,
State Arboretum of
Virginia

Buxus microphylla
'Grace Hendrick
Phillips',
0.75 ft. h x 2 ft. w,
age 17 years ,
State Arboretum
of Virginia.
Photo by P. D. Larson

Buxus microphylla 'John Baldwin', 6¼ ft. h x 4½ ft. w, age 27 years, State Arboretum of Virginia. Photo by C. A. Flanagan

Buxus microphylla (with C, Sacchi), 12 ft. h x 12½ ft. w, age 33 years, State Arboretum of Virginia. Photo by C. A. Flanagan

Buxus sempervirens 'Vardar Valley', 5½ ft. h x 12 ft. w, age 38 years, State Arboretum of Virginia. Photo by C. A. Flanagan

Buxus sempervirens
'Glauca',
9½ ft. h x 12 ft. w,
age 32 years,
State Arboretum of
Virginia.
Photo by P. D. Larson

Buxus sempervirens
'Elegantissima'
prostrate form,
2¾ ft. h x 2½ ft. w,
age 12 years,
State Arboretum
of Virginia.
Photo by
C. A. Flanagan

Buxus sempervirens
'Butterworth',
7½ h x 7 ft. w,
age 25 years,
State Arboretum
of Virginia.
Photo by
P. D. Larson

Buxus sempervirens 'Denmark'

Size (25 yrs.):	Large – 9 to 10 feet high and 8 to 9 feet wide.
Natural Form:	Pyramidal with stiffly upright branches and somewhat loose and open habit.
Annual Growth:	Fast – $4^1/_2$ to 5 inches in height and 4 to $4^1/_2$ inches in width.
Leaf Color:	Medium yellow-green; prone to bronzing in the winter when sited in the direct sun.
Leaf Shape:	Rotund and slightly revolute; obtuse and retuse tip; cuneate base.
Leaf Size:	Large – $1^1/_8$ to $1^3/_{16}$ inches long and $5/_8$ to $^{11}/_{16}$ inch wide.
Leaf Surface:	Glossy and smooth.
Internodal Length:	Medium – $^7/_{16}$ to $^3/_4$ inch.
Flowering Habit:	Sparse flowering and sparse fruiting.
Hardiness:	Zones 5(protected) to 8.
Plant Use:	Specimen, foundation planting, grouping for background and area separations, hedgings.

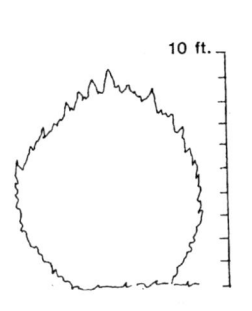

Registration: Mary A. Gamble and P.D. Larson in *The Boxwood Bulletin*, Vol. 28(2):28, Oct. 1988.
History: Mr. Bagby, a plant propagator at Gray Summit, Mo., received cuttings from Neils Alfred Paludan (an exchange student at the Missouri Botanical Garden) of Hellerup, Denmark in 1953. Efforts to determine the parent clone of these original cuttings have been unsuccessful.
Bibliography:
The Boxwood Bulletin, Vol. 28(2):28, Oct. 1988/28(3):42, Jan. 1989.
Known Locations: U.S. National Arboretum, State Arboretum of Virginia.
Additional Information:
Culture and Care: Demonstrates no special cultural requirements.
Pests and Diseases: Indicates resistance to leaf miner, psyllid and mites in the more humid climates; no serious diseases.
Not available in the commercial nursery trade.

Buxus sempervirens 'Edgar Anderson'

Size (25 yrs.):	Medium – 4 to 5 feet high and 3½ to 4 feet wide. A 31-year-old specimen measures 9 feet high and 9 feet wide.
Natural Form:	Pyramidal – irregular outline, dense foliage.
Annual Growth:	Medium – 1¾ to 2 inches in height and 1 to 1½ inches in width.
Leaf Color:	Medium green.
Leaf Shape:	Elliptic – slightly revolute, some ovate; acute tip; cuneate base.
Leaf Size:	Small – ⁵/₈ to ¾ inch long and ³/₁₆ to ¼ inch wide.
Leaf Surface:	Glabrous and smooth.
Internodal Length:	Short – ³/₁₆ to ¼ inch.
Flowering Habit:	Not observed.
Hardiness:	Zones 5(protected) to 8.
Plant Use:	Specimen, foundation planting, grouping for background and area separations, hedgings.

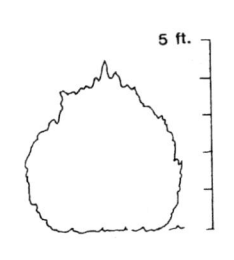

Registration: Mary Gamble in *The Boxwood Bulletin,* Vol. 13(2):26-28, Oct. 1973.
History: Cuttings were collected by Dr. Edgar Anderson from Romania. They were propagated at Arnold Arboretum, Jamaica Plain, Mass., in 1935. The accession book at Arnold Arboretum labeled the original cuttings as "351-35 *B. semp.* #2 E.A. Bucharesk, Roumania, 58 cuttings."
Bibliography:
The Boxwood Society of the Midwest Bulletin, April 1986.
The Boxwood Bulletin, Vol. 13(2):26-28, Oct. 1973/14(4):58-59, April 1975/16(1):14, July 1976/20(3):43,45, Jan. 1981/21(4):65, April 1982/24(2):44-46, Oct. 1984.
Known Locations: Missouri Botanical Garden, Washington Park Arboretum, U.S. National Arboretum, State Arboretum of Virginia.
Additional Information:
Culture and Care: Tolerates being sited in some direct sun. Demonstrates no special cultural requirements.
Pests and Diseases: Resistance to leaf miner and psyllid.
Not available in the commercial nursery trade.

Buxus sempervirens 'Elegantissima'

Size (25 yrs.):	Large – 7 to 8 feet high and 6½ to 7 feet wide. A 40-year-old specimen measures 12 feet high and 6 feet wide.
Natural Form:	Pyramidal – loose open habit.
Annual Growth:	Fast – 3½ to 4 inches in height and 3 to 4 inches in width.
Leaf Color:	Medium green with variegations of gold at the margins.
Leaf Shape:	Lanceolate – variegated, distorted, revolute; obtuse tip; cuneate base.
Leaf Size:	Medium – ¾ to 1 inch long and ⅜ to ⅝ inch wide.
Leaf Surface:	Glabrous and smooth.
Internodal Length:	Long – ⅜ to 7/16 inch.
Flowering Habit:	Not observed.
Hardiness:	Zones 6 to 9.
Plant Use:	Specimen.

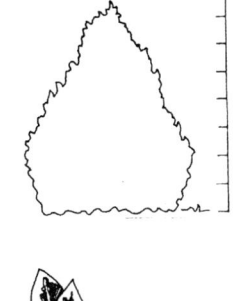

Registration: Hort. ex K. Koch in *Dendrologie* v. 2, ot. 2:471:1872. (Synonym 'Elegantissima Variegata'). Catalog, Charles Dietriche, Angers, France, 1892.
History: Most likely originated as an open-pollinated seedling or mutation somewhere in Europe.
Bibliography:
Bailey, L.H. *Hortus Third*, 1976.
Dallimore, W. *Holly, Yew and Box*, 1908.
Dirr, M.A. *Manual of Woody Landscape Plants*, 4th Ed., 1990.
Wright, M. *The Complete Handbook of Garden Plants*, 1984.
The Boxwood Bulletin, Vol. 2(3):37, Jan. 1963/8(3):45, Jan. 1969/9(1):7, July 1969/ 11(2):31, Oct. 1971/17(2):25, Oct. 1977/20(4):80, April 1981/27(3):65, Jan. 1988.
Known Locations: Buzzards Bay Garden Club, Royal Botanic Gardens, Edinburgh, Hillier's Arboretum, Royal Botanic Gardens, Kew, Oxford Botanic Gardens, Washington Park Arboretum, U.S. National Arboretum, State Arboretum of Virginia.
Additional Information:
Culture and Care: Tolerates siting in direct sun quite well. Breaks dormancy late.
Pests and Diseases: Resistence to leaf miner, psyllid and mites.
Propagation: Difficult and quite slow to root. Take cuttings when new spring growth has hardened, taking the cuttings slightly below a node; bruise the stem or slit each side of the stem; use an IBA powder dip.
Available in the commercial nursery trade.

Buxus sempervirens 'Fastigiata'

Size (25 yrs.):	Large – 9 to 10 feet high and 2 to 2½ feet wide. A 40-year-old specimen measures 12 feet high and 5 feet wide.
Natural Form:	Conical with stiffly upright branches, dense foliage.
Annual Growth:	Fast – 3½ to 3¾ inches in height and ¾ to 1 inch in width.
Leaf Color:	Dark green.
Leaf Shape:	Lanceolate – slightly revolute; acute tip; cuneate base.
Leaf Size:	Large – ⅞ to 1⅛ inches long and ⅜ to ½ inch wide.
Leaf Surface:	Glossy and smooth.
Internodal Length:	Long – ½ to ¹¹⁄₁₆ inch.
Flowering Habit:	Sparse flowering and fruiting.
Hardiness:	Zones 6 to 8.
Plant Use:	Specimen, grouping for background and area separations, hedgings, allees.

Registration: F. Meyer. *Plant Explorations*, ARS 34-9:91, 1959. Sometimes erroneously referred to as 'Hardwickensis Fastigiata.'
History: Most likely originated as an open-pollinated seedling somewhere in Europe.
Bibliography:
Bailey, L.H. *Hortus Third*, 1976.
Flint, H.L. *Horticulture,* March 1987.
The Boxwood Bulletin, Vol. 20(4):80, April 1981/21(4):65, April 1982/27(3):65, Jan. 1988.
Known Locations: Buzzards Bay Garden Club, Washington Park Arboretum, U.S. National Arboretum, State Arboretum of Virginia.
Additional Information:
Culture and Care: This is an excellent plant and another of my favorites. Tolerates siting in direct sun quite well. Breaks dormancy late.
Pests and Diseases: Resistance to leaf miner and mites. Indicates some attraction for psyllids.
Available in the commercial nursery trade.

Buxus sempervirens 'Flora Place'

Size (25 yrs.):	Large – 6 to 7 feet high and 4 to 5 feet wide.
Natural Form:	Pyramidal, open and rangey.
Annual Growth:	Medium – 3 to 3$^1/_2$ inches in height and 2 to 2$^1/_2$ inches in width.
Leaf Color:	Dark green with yellow undertones.
Leaf Shape:	Lanceolate to elliptic and slightly revolute; acute tip; cuneate base.
Leaf Size:	Large – $^7/_8$ to 1$^1/_8$ inch long and $^5/_{16}$ to $^5/_8$ inch wide.
Leaf Surface:	Glabrous and smooth.
Internodal Length:	Long – $^7/_{16}$ to $^5/_8$ inch.
Flowering Habit:	Sparse flowering and sparse fruiting.
Hardiness:	Zones 5 to 8.
Plant Use:	Specimen, grouping for background and area separations.

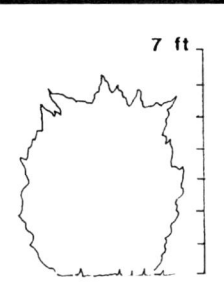

Registration: Mary A. Gamble for the Boxwood Society of the Midwest in *The Boxwood Bulletin,* Vol. 28(4):59, April 1989.
History: Seed sent to Dr. Edgar Anderson from the Yugoslavian Forestry Service in 1935. The seed was propagated at the Missouri Botanical Garden and tested at Gray Summit, Mo., and Kingsville Nurseries, Kingsville, Md. It was named after a city street in St. Louis, Mo., where clones were also grown by Dr. Edgar Anderson and Paul A. Kohl.
Bibliography:
The Boxwood Bulletin, Vol. 28(4):59-60, April 1989.
Known Locations: Missouri Botanical Garden, State Arboretum of Virginia.
Additional Information:
Culture and Care: Demonstrates no special cultural requirements.
Pests and Diseases: Indicates resistance to leaf miner, psyllid and mites in the more humid climates; no serious diseases.
Not available in the commercial nursery trade.

Buxus sempervirens 'Fortunei Rotundifolia'

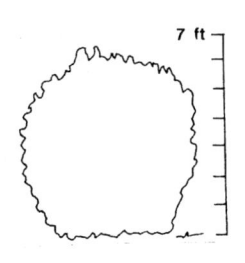

Size (25 yrs.):	Large – 6 to 7 feet high and 7 to 8 feet wide. A 29-year-old specimen measures 7 feet high and 8 feet wide.
Natural Form:	Pyramidal, billowy and somewhat loose and open habit.
Annual Growth:	Medium – 2½ to 3 inches in height and 3 to 3½ inches in width.
Leaf Color:	Medium yellow-green, prone to bronze in winter when sited in the direct sun.
Leaf Shape:	Rotund tending toward lanceolate; obtuse tip; cuneate base.
Leaf Size:	Medium – ½ to ¹³/₁₆ inch long and ⁵/₁₆ to ⁷/₁₆ inch wide.
Leaf Surface:	Glossy and smooth.
Internodal Length:	Long – ⁷/₁₆ to ⅝ inch.
Flowering Habit:	Moderate flowering and moderate fruiting.
Hardiness:	Zones 6 to 8.
Plant Use:	Specimen, grouping for background and area separations.

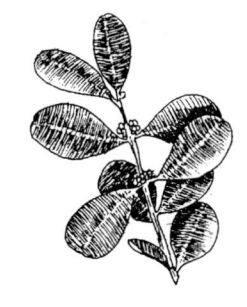

Registration: Not registered.
History: Appeared in the 1961 inventory of Royal Botanic Gardens, Kew, England.
Bibliography:
The Boxwood Society of the Midwest Bulletin, Vol. 12(1):7, July 1988.
Known Locations: Missouri Botanical Garden, U.S. National Arboretum, State Arboretum of Virginia.
Additional Information:
Culture and Care: Demonstrates no special cultural requirements.
Pests and Diseases: Indicates resistance to leaf miner, psyllid and mites in the more humid climates; no serious diseases.
Not available in the commercial nursery trade.

Buxus sempervirens 'Glauca'

Size (25 yrs.):	Large – 6½ to 7 feet high and 5½ to 6 feet wide.
Natural Form:	Pyramidal, billowy with upright growth and loose open habit.
Annual Growth:	Medium – 2½ to 3 inches in height and 2 to 3 inches in width.
Leaf Color:	Dark green with a bluish hue.
Leaf Shape:	Lanceolate and slightly revolute; acute tip tending toward obtuse; cuneate base.
Leaf Size:	Large – 1 to 1¹/₆ inches long and ³/₈ to ⁷/₁₆ inch wide.
Leaf Surface:	Glaucous – covered with a "bloom" somewhat similar to that of a peach.
Internodal Length:	Long – ⁷/₁₆ to ⅝ inch.
Flowering Habit:	Floriferous, heavy fruit set.
Hardiness:	Zones 6 to 8.
Plant Use:	Specimen, hedging, grouping for background and area separations.

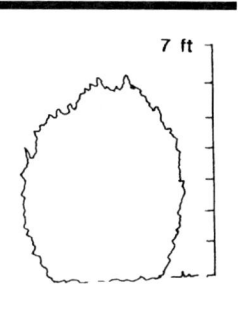

Registration: G. Kirchner in Petzold and Kirchner, *Arboretum Muscaviense* 194, 1864. Sometimes erroneously referred to as 'Macrophylla Glauca.'
History: Most likely originated as an open-pollinated seedling somewhere in Europe.
Bibliography:
Bailey, L.H. *Hortus Third*, 1976.
Hotte, A.C. *The Book of Shrubs*, 1931.
Krussman, G. *Manual of Cultivated Broad-leaved Trees and Shrubs*, Vol. I, A-D, 1986.
The Boxwood Bulletin, Vol. 9(1):7, July 1969.
Known Locations: Brooklyn Botanic Garden; Royal Botanic Gardens, Edinburgh, Scotland; Royal Botanic Gardens, Kew, England; U.S. National Arboretum; State Arboretum of Virginia.
Additional Information:
Culture and Care: Demonstrates no special cultural requirements.
Pests and Diseases: Indicates resistance to leaf miner, psyllid and mites in the more humid climates; no serious diseases.
Available in the commercial nursery trade of North America and Europe.

Selected Species, Varieties, Cultivars

Buxus sempervirens 'Graham Blandy'

Size (25 yrs.):	Large – 10 to 12 feet high and 1 to 1½ feet wide.
Natural Form:	Columnar – single straight vertical main stem, stiff erect branches.
Annual Growth:	Fast – 4½ to 5 inches in height and ½ to ¾ inch in width.
Leaf Color:	Medium green.
Leaf Shape:	Lanceolate to elliptic, some rotund; obtuse tip; cuneate base.
Leaf Size:	Large – ¾ to ⅞ inch long and ⅜ to ½ inch wide.
Leaf Surface:	Glossy and smooth.
Internodal Length:	Medium – ¼ to 5/16 inch.
Flowering Habit:	Not observed.
Hardiness:	Zones 6 to 8.
Plant Use:	Specimen, grouping for background and area separations, allees.

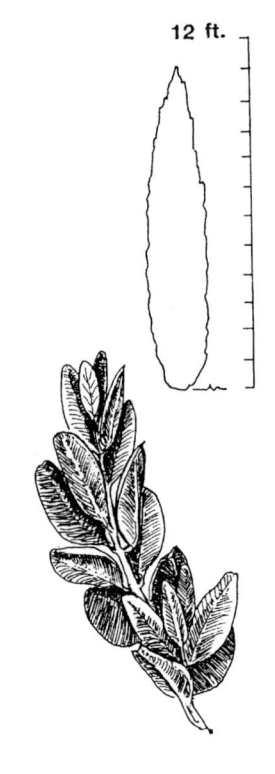

Registration: T. Ewert, director of Blandy Experimental Farm, Boyce, Va., in *The Boxwood Bulletin*, Vol. 25(1):8, July 1985. Formerly referred to as 'BEF-35.'

History: Discovered in 1949 at Blandy Experimental Farm. Named to honor Graham F. Blandy who in 1926 bequeathed more than 700 acres of his property to the University of Virginia to establish Blandy Experimental Farm. Dr. Orland E. White, the first director of Blandy, promoted plant systematic research in many families, including the Buxaceae.

Bibliography:
Dirr, M.A. *Manual of Woody Landscape Plants*, 4th Ed., 1990.
Flint, H.L. *Horticulture*, March 1987.
The Boxwood Bulletin, Vol. 8(4):63,64, April 1969/25(1):Cover,8, July 1985/27(3):65, Jan. 1988.

Known Locations: Buzzards Bay Garden Club, U.S. National Arboretum, State Arboretum of Virginia.

Additional Information:
Culture and Care: Tolerates siting in direct sun quite well but should be sited in an area protected from winter winds to preclude burn and bronzing. Due to the plant's fairly rapid growth, the outer stems are prone to droop somewhat from heavy rains; judiciously prune half the new spring growth of the outer stems to stiffen them and the problem will disappear. Breaks dormancy early.
Pests and Diseases: Prone to leaf miner and psyllid infestation.
Available in the commercial nursery trade.

Buxus sempervirens 'Handsworthiensis'

Size (25 yrs.):	Large – 8 to 9 feet high and 8 to 9 feet wide. A 29-year-old specimen measures 10 feet high and 10 feet wide.
Natural Form:	Pyramidal with a loose upright habit.
Annual Growth:	Fast – 4 to 4^1/$_2$ inches in height and 4 to 4^1/$_2$ inches in width.
Leaf Color:	Dark green with orange-tinged twigs particularly in winter.
Leaf Shape:	Lanceolate tending toward ovate and slightly revolute; acute to rounded tip with some being retuse; cuneate base.
Leaf Size:	Large – ^{3}/$_4$ to 1^1/$_{16}$ inches long and ^{3}/$_8$ to ^{11}/$_{16}$ inch wide.
Leaf Surface:	Glossy and smooth.
Internodal Length:	Medium – ^{1}/$_4$ to ^{3}/$_8$ inch.
Flowering Habit:	Floriferous and heavy fruiting.
Hardiness:	Zones 6 to 8.
Plant Use:	Specimen.

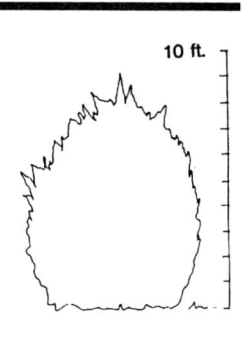

Registration: Fisher ex Henry in Elses and Henry, *Trees of Britain and Ireland*, no. 7:1725. date: 1913. Invalid names: 'Navicularis,' 'Suffruticosa Navicularis.'
History: Originated and named by Fisher Son & Sibrag, Handsworth Nursery located near Sheffield, England.
Bibliography:
Bailey, L.H. *Hortus Third*, 1976.
Dirr, M.A. *Manual of Woody Landscape Plants*, 4th Ed., 1990.
Everett, T.H. *The New York Botanical Garden Ill. Encyc. of Hort.*, Vol. 2, 1981.
Flint, H.L. *Horticulture,* March 1987.
Krussman, G. *Manual of Cultivated Broad-leaved Trees and Shrubs*, Vol. I, A-D, 1986.
Wright, M. *The Complete Handbook of Garden Plants*, 1984.
Wyman, D. *Wyman's Gardening Encyclopedia*, New Expanded 2nd Ed., 1986.
The Boxwood Bulletin, Vol. 8(4):60, April 1969/9(1):7, July 1969/16(1):15, July 1976/ 20(4):80, April 1981/21(3):45, Jan. 1982/21(4):67, April 1982.
Known Locations: Buzzards Bay Garden Club, Hillier's Arboretum, Royal Botanic Gardens, Kew;` Washington Park Arboretum, U.S. National Arboretum, State Arboretum of Virginia.

Additional Information:
Culture and Care: Demonstrates no special cultural requirements.
Pests and Diseases: Indicates resistance to leaf miner, psyllid and mites in the more humid climates; no serious diseases.
Available in the commercial nursery trade.

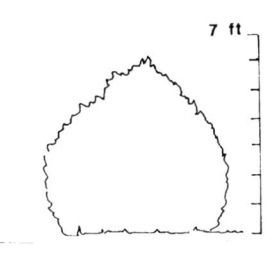

Buxus sempervirens 'Handsworthii'

Size (25 yrs.):	Large – 7 to 8 feet high and 6½ to 7 feet wide. A 57-year-old specimen measures 8 feet high and 7 feet wide.
Natural Form:	Pyramidal, stiffly upright with a loose and open habit.
Annual Growth:	Fast – 3 to 3½ inches in height and 2½ and 3 inches in width.
Leaf Color:	Dark green.
Leaf Shape:	Lanceolate tending toward rotund and slightly revolute; acute tip with some being slightly retuse; cuneate base.
Leaf Size:	Large – 1 to 1⅛ inches long and ⅜ to ½ inch wide.
Leaf Surface:	Matte and smooth.
Internodal Length:	Short – ¼ to 5/16 inch.
Flowering Habit:	Moderate flowering and sparse fruiting.
Hardiness:	Zones 6 to 8.
Plant Use:	Specimen.

Registration: Hort. ex K. Koch in *Dendrologie*, v.2. pt.2:476, 1872.
History: Most likely originated as an open-pollinated seedling somewhere in Europe.
Bibliography:
Bailey, L.H. *Hortus Third*, 1976.
Dallimore, W. *Holly, Yew and Box*, 1908.
Hotte, A.C. *The Book of Shrubs*, 1931.
The Boxwood Bulletin, Vol. 2(3):37, Jan. 1963/8(4):60, April 1969.
Known Locations: College of William and Mary, State Arboretum of Virginia.
Additional Information:
Culture and Care: Demonstrates no special cultural requirements.
Pests and Diseases: Indicates resistance to psyllid; no serious diseases.
Available in the commercial nursery trade.

Selected Species, Varieties, Cultivars

Buxus sempervirens 'Hardwickensis'

Size (25 yrs.):	Medium – 4 to 4½ feet high and 3 to 3½ feet wide.
Natural Form:	Unusual – somewhat columnar to vase-shaped, stiff upright candle-like branches.
Annual Growth:	Medium – 2 to 2½ inches in height and 1½ to 2 inches in width.
Leaf Color:	Dark green.
Leaf Shape:	Rotund – slightly revolute; obtuse tip, some retuse; cuneate base. Unusual circular leaf arrangement on branches.
Leaf Size:	Small – ³/₈ to ⁵/₈ inch long and ³/₈ to ⁵/₁₆ inch wide.
Leaf Surface:	Glabrous and smooth.
Internodal Length:	Medium – ¼ to ⁵/₁₆ inch.
Flowering Habit:	Not observed.
Hardiness:	Zones 6 to 8.
Plant Use:	Specimen.

Registration: Beissner, Schelle and Zabel in *Handbuch der Laubholz-Benennung*, 283, 1903.
History: Most likely originated as an open-pollinated seedling somewhere in Europe.
Bibliography: Not documented
Known Locations: College of William and Mary, Hillier's Arboretum, Washington Park Arboretum, U.S. National Arboretum, State Arboretum of Virginia.
Additional Information:
Culture and Care: Prefers dappled shade but will tolerate being sited in some direct sun.
Pests and Diseases: Resistance to leaf miner, psyllid and mites.
Available in the commercial nursery trade.

Buxus sempervirens 'Heinrich Bruns'

Size (25 yrs.):	Medium – 3½ to 4 feet in height, narrower than wide.
Natural Form:	Unusual, particularly as it relates to leaf shape, form and general appearance.
Annual Growth:	Medium – 1½ to 2 inches in both height and width.
Leaf Color:	Dark green.
Leaf Shape:	Ovate tending toward rotund, revolute; obtuse tip, some retuse; cuneate base. The leaves appear to be whorled and somewhat bullate on the branchlets.
Leaf Size:	Medium – ³⁄₈ to ¾ inch long and ¼ to ³⁄₈ inch wide.
Leaf Surface:	Glabrous and smooth.
Internodal Length:	Medium – ¼ to ³⁄₈ inch.
Flowering Habit:	Not observed.
Hardiness:	Zones 6 to 8.
Plant Use:	Specimen.

Registration: F.G. Meyer in New Cultivars of Woody Ornamentals From Europe. *Baileya* 9(4):129, 1961.
History: Originated from the nursery of Heinrich Bruns, Westerstede, Oldenberg, Germany. Introduced in North America by the U.S. Department of Agriculture as PI 260383.
Bibliography:
The Boxwood Bulletin, Vol. 7(1):1, July 1967.
Known Locations: Washington Park Arboretum, U.S. National Arboretum, State Arboretum of Virginia.
Additional Information:
Culture and Care: Demonstrates no special cultural requirements.
Pests and Diseases: Indicates resistance to leaf miner, psyllid and mites in the more humid climates; no serious diseases.
Not available in the commercial nursery trade.

Buxus sempervirens 'Henry Shaw'

Size (25 yrs.):	Medium – 5 to 6 feet high and 4 to 5 feet wide.
Natural Form:	Pyramidal – billowy.
Annual Growth:	Medium – 2 to 2½ inches in height and 1½ to 2 inches in width.
Leaf Color:	Medium green with slight yellow undertone.
Leaf Shape:	Lanceolate – slightly revolute; acute tip, some retuse; cuneate base.
Leaf Size:	Medium – ½ to ⅞ inch long and 3/16 to 7/16 inch wide.
Leaf Surface:	Glabrous and smooth.
Internodal Length:	Medium – ¼ to 5/16 inch.
Flowering Habit:	Moderate flowering; sparse fruiting.
Hardiness:	Zones 5 to 8.
Plant Use:	Specimen, foundation planting, grouping for background and area separations, hedgings.

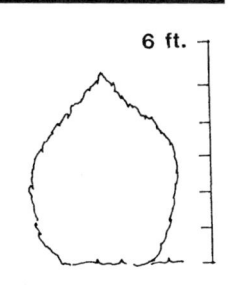

Registration: Mary A. Gamble in *The Boxwood Bulletin*, Vol. 25(2):47, Oct. 1985. Sometimes erroneously referred to as Field Row, which was a convenience name. It was also carried by number 752075 at the Missouri Botanical Garden.

History: Collected by Dr. Edgar Anderson from the Balkans. I believe this cultivar to be a second generation of test plants and not part of the earlier seed-grown K-series. Plant was named for Henry Shaw, an Englishman, who came to St. Louis, Mo., in 1819 and established a thriving business. Established the Missouri Botanical Garden, which he left to the city of St. Louis upon his death in 1889.

Bibliography:
Missouri Botanical Garden Bulletin, Vol. LXXVI(3):8-10, May-June 1988.
The Boxwood Society of the Midwest Bulletin, April 1986.
The Boxwood Bulletin, Vol. 25(2):43-47, Oct. 1985.

Known Locations: Buzzards Bay Garden Club, Missouri Botanical Garden, U.S. National Arboretum, State Arboretum of Virginia.

Additional Information:
Culture and Care: Tolerates siting in direct sun quite well. Breaks dormancy late.
Pests and Diseases: Resistance to leaf miner, psyllid and mites.
Not available in the commercial nursery trade.

Buxus sempervirens 'Hermann von Schrenk'

Size (25 yrs.):	Large – 7 to 8 feet high and 9 to 10 feet wide. A 36-year-old specimen measures 8 feet high and 10½ feet wide.
Natural Form:	Pyramidal, slightly irregular habit.
Annual Growth:	Medium – 3 to 3½ inches in height and 4 to 5 inches in width.
Leaf Color:	Medium green.
Leaf Shape:	Elliptic to lanceolate and slightly revolute; acute tip; cuneate base.
Leaf Size:	Medium – ½ to ¹¹/₁₆ inch long and ³/₁₆ to ⁵/₁₆ inch wide.
Leaf Surface:	Glabrous and smooth.
Internodal Length:	Short – ³/₁₆ to ¼ inch.
Flowering Habit:	Not observed.
Hardiness:	Zones 5 to 8.
Plant Use:	Specimen, foundation planting, grouping for background and area separations, hedgings.

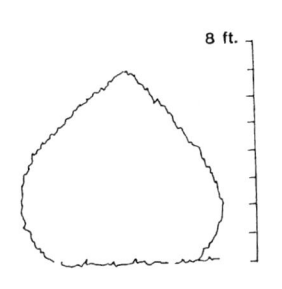

Registration: Mary A. Gamble in *The Boxwood Bulletin*, Vol. 14(2):31, Oct. 1974.
History: Originated in Charlottesville, Va., and was brought to the Missouri Botanical Garden in the early 1930s by Dr. Hermann von Schrenk, a scientist and plant pathologist associated with the botanical garden.
Bibliography:
The Boxwood Society of the Midwest Bulletin, April 1986.
The Boxwood Bulletin, Vol. 14(2):Cover,31,i.b.c., Oct. 1974/ 20(3):41,42,44, Jan. 1981/ 21(4):65, April 1982/23(1):19-20, July 1983/25(4):112, April 1986/27(2):37, Oct. 1987.
Known Locations: Buzzards Bay Garden Club, Missouri Botanical Garden, Secrest Arboretum, U.S. National Arboretum, State Arboretum of Virginia.
Additional Information:
Culture and Care: Demonstrates no special cultural requirements.
Pests and Diseases: Indicates resistance to leaf miner, psyllid and mites in the more humid climates; no serious diseases.
Not available in the commercial nursery trade.

Buxus sempervirens 'Holland'

Size (25 yrs.):	Medium – 5$\frac{1}{2}$ to 6 feet high and 4 to 4$\frac{1}{2}$ feet wide. A 40-year-old specimen measures 12 feet high and 8$\frac{1}{2}$ feet wide.
Natural Form:	Conical with loose open habit.
Annual Growth:	Medium – 3 to 3$\frac{1}{2}$ inches in height and 2 to 2$\frac{1}{2}$ inches in width.
Leaf Color:	Medium green.
Leaf Shape:	Ovate; retuse tip; cuneate base.
Leaf Size:	Medium – $\frac{7}{8}$ to 1 inch long and $\frac{5}{16}$ to $\frac{3}{8}$ inch wide.
Leaf Surface:	Glabrous and smooth.
Internodal Length:	Medium – $\frac{1}{4}$ to $\frac{3}{8}$ inch.
Flowering Habit:	Not observed.
Hardiness:	Zones 6 to 8.
Plant Use:	Specimen, grouping for background and area separations, hedgings, allees.

Registration: Not registered.
History: Believed to have originated as an open-pollinated seedling somewhere in Michigan. It was listed in the Weller Nursery Company catalog in the early 1940s.
Bibliography: Not documented
Known Locations: U.S. National Arboretum, State Arboretum of Virginia.
Additional Information:
Culture and Care: Demonstrates no special cultural requirements.
Pests and Diseases: Indicates resistance to leaf miner, psyllid and mites in the more humid climates; no serious diseases.
Not available in the commercial nursery trade.

Buxus sempervirens 'Hood'

Size (25 yrs.):	Medium – 3½ to 4 feet high and 3 to 3½ feet wide.
Natural Form:	Pyramidal and billowy.
Annual Growth:	Slow – 1 to 1½ inches in height and ¾ to 1 inch in width.
Leaf Color:	Dark green.
Leaf Shape:	Lanceolate and quite uniform; acute tip; cuneate base.
Leaf Size:	Medium – ½ to ¾ inch long and ³⁄₁₆ to ¼ inch wide.
Leaf Surface:	Glabrous and smooth.
Internodal Length:	Medium – ¼ to ⁵⁄₁₆ inch.
Flowering Habit:	Not observed.
Hardiness:	Zones 5 to 8.
Plant Use:	Specimen, foundation planting, grouping for background and area separations, edgings, hedgings.

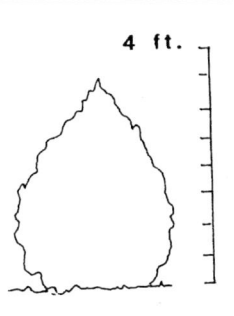

Registration: The Boxwood Society of the Midwest in *The Boxwood Bulletin*, Vol. 26(3):64-67, Jan. 1987.
History: Originally came from the cemetery that adjoins the St. Augustine Catholic Church in Kelso, Scott County, Mo. Mrs. Ethel B. Hood of Flat River, Mo., provided cuttings to The Boxwood Society of the Midwest in 1972 from 20-year-old plants growing in her backyard.
Bibliography:
The Boxwood Society of the Midwest Bulletin, April 1986.
The Boxwood Bulletin, Vol. 26(3):64-67, Jan. 1987.
Known Locations: Missouri Botanical Garden.
Additional Information:
Culture and Care: Demonstrates no special cultural requirements.
Pests and Diseases: Indicates resistance to leaf miner, psyllid and mites in the more humid climates; no serious diseases.
Available in the commercial nursery trade.

Buxus sempervirens 'Inglis'

Size (25 yrs.):	Large – 9 to 10 feet high and 8 to 9 feet wide.	10 ft
Natural Form:	Pyramidal – billowy, compact.	
Annual Growth:	Fast – 4 to 4$^{1}/_{2}$ inches in height and 3$^{1}/_{2}$ to 4 inches in width.	
Leaf Color:	Medium green with a bluish hue.	
Leaf Shape:	Elliptic to lanceolate, slightly revolute; acute tip; cuneate base.	
Leaf Size:	Small – $^{7}/_{16}$ to $^{11}/_{16}$ inch long and $^{3}/_{16}$ to $^{5}/_{16}$ inch wide.	
Leaf Surface:	Glabrous and smooth.	
Internodal Length:	Short – $^{3}/_{16}$ to $^{1}/_{4}$ inch.	
Flowering Habit:	Floriferous; heavy fruiting.	
Hardiness:	Zones 5 to 8.	
Plant Use:	Specimen, grouping for background and area separations, hedgings.	

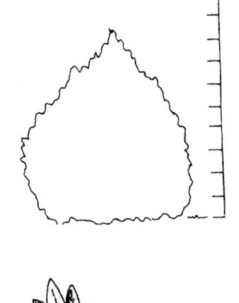

Registration: D. Wyman in *Arnoldia*, 17(11):65, 1957.
History: Members of the Federated Garden Clubs of Michigan discovered the plant in a member's garden and learned that a clone had been propagated from boxwood sprigs which had decorated a gift box of fruit sent from New York in the early 1930s. Attempts to trace the ancestry of the plant have failed. The cultivar was named for Elizabeth H. Inglis.
Bibliography:
Dirr, M.A. *Manual of Woody Landscape Plants*, 4th Ed., 1990.
Gamble, M.A. *Flower and Garden,* March 1988.
Missouri Botanical Garden Bulletin, Vol. LXXVI(3):8-10, May-June 1988.
The Boxwood Society of the Midwest Bulletin, April 1986.
The Boxwood Bulletin, Vol. 2(4):44, April 1963/3(1):1, July 1963/20(3):42,44, Jan. 1981/20(4):80, April 1981/29(3):50-51, Jan. 1989.
Known Locations: Arnold Arboretum, Brooklyn Botanic Garden, Buzzards Bay Garden Club, Missouri Botanical Garden, U.S. National Arboretum, State Arboretum of Virginia.
Additional Information:
Culture and Care: Tolerates siting in direct sun quite well. Breaks dormancy late.
Pests and Diseases: Resistance to leaf miner, psyllid and mites.
Available in the commercial nursery trade.

Buxus sempervirens 'Ipek'

Size (25 yrs.):	Large – 10 to 11 feet high and 6 to 7 feet wide.
Natural Form:	Pyramidal, somewhat billowy with loose and open habit.
Annual Growth:	Fast – 5 to 5^1/$_2$ inches in height and 3 to 4 inches in width.
Leaf Color:	Dark green with bluish hue.
Leaf Shape:	Lanceolate; acute tip; cuneate base.
Leaf Size:	Large – ^{3}/$_4$ to 1 inch long and ^{1}/$_4$ to ^{7}/$_{16}$ inch wide.
Leaf Surface:	Glossy and smooth.
Internodal Length:	Medium – ^{5}/$_{16}$ to ^{3}/$_8$ inch.
Flowering Habit:	Sparse flowering and sparse fruiting.
Hardiness:	Zones 5 to 8.
Plant Use:	Specimen, grouping for background and area separations.

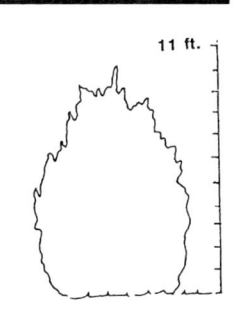

Registration: Not registered.
History: Grown by Dr. Edgar Anderson from seed sent to him by the Yugoslavian Forestry Service in 1935. Propagated at the Missouri Botanical Garden, tested at Gray Summit, Mo., and Kingsville Nurseries, Kingsville, Md. Named by Dr. Anderson for a town in Yugoslavia, now written Pec.
Bibliography:
The Boxwood Society of the Midwest Bulletin, Vol. XII(1):7, July 1988.
The Boxwood Bulletin, Vol. 24(2):51, Oct. 1984.
Known Locations: Missouri Botanical Garden. U.S. National Arboretum, State Arboretum of Virginia.
Additional Information:
Culture and Care: Demonstrates no special cultural requirements.
Pests and Diseases: Indicates resistance to leaf miner, psyllid and mites in the more humid climates; no serious diseases.
Not available in the commercial nursery trade.

Selected Species, Varieties, Cultivars

Buxus sempervirens 'Joe Gable'

Size (25 yrs.):	Large – 6 feet in height and 6 to 7 feet in width.
Natural Form:	Pyramidal, open and loose habit.
Annual Growth:	Medium – 2 to 2½ inches in height and 2½ to 3 inches in width.
Leaf Color:	Medium green.
Leaf Shape:	Elliptic; acute tip tending toward obtuse; cuneate base; produces a wilowy effect.
Leaf Size:	Medium – ¾ to 1 inch long and 3/16 to 3/8 inch wide.
Leaf Surface:	Glabrous and smooth, tending toward glossy.
Internodal Length:	Medium – 5/16 to 3/8 inch.
Flowering Habit:	Not observed.
Hardiness:	Zones 6 to 8.
Plant Use:	Specimen.

Registration: Not registered.
History: The plant is believed to have originated as an open-pollinated seedling and was discovered by Joe Gable of Stewartstown, Pa., who sent it to Henry Hohman of Kingsville Nurseries, Kingsville, Md., for further propagation, testing, naming and release. The following appeared in the 1946 catalog of Kingsville Nurseries: "*B. semp.* 'Joe Gable' – dark green leaves, holding color in very cold weather. Growth fast and strong. It is apparent that this box will develop to quite a large size."
Bibliography: Not documented
Known Locations: Washington Park Arboretum, U.S. National Arboretum, State Arboretum of Virginia.
Additional Information:
Culture and Care: Demonstrates no special cultural requirements.
Pests and Diseases: Indicates resistance to leaf miner, psyllid and mites in the more humid climates; no serious diseases.
Not available in the commercial nursery trade.

Buxus sempervirens 'Joy'

Size (25 yrs.):	Large – 9 to 10 feet high and 7 to 8 feet in width.
Natural Form:	Pyramidal tending toward conical, with open habit; tendency to have some candle-like branches in earlier years.
Annual Growth:	Fast – 4 to 5 inches in height and 3 to 4 inches in width.
Leaf Color:	Medium green.
Leaf Shape:	Lanceolate – revolute; acute tip; cuneate base.
Leaf Size:	Medium – $1/2$ to $13/16$ inch long and $3/16$ to $5/16$ inch wide.
Leaf Surface:	Glabrous and smooth.
Internodal Length:	Medium – $5/16$ to $3/8$ inch.
Flowering Habit:	Not observed.
Hardiness:	Zones 5 to 8.
Plant Use:	Specimen, foundation planting, grouping for background and area separations, edgings, hedgings.

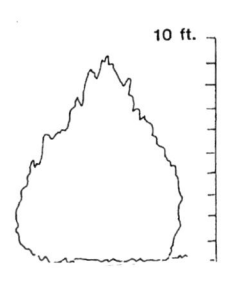

Registration: Mary Gamble in *The Boxwood Bulletin*, Vol. 24(1):12-13, July 1984.
History: The original plant is not known. Cuttings came from the garden of Mrs. Marion Rombauer Becker of Ohio in 1970. These plants were then 20 years old and had come from Michigan. They were carried under the convenience name 'Becker' by the Boxwood Society of the Midwest. The name 'Joy' honors the late Marion Rombauer, co-editor of *The Joy of Cooking*.
Bibliography:
Missouri Botanical Garden Bulletin, Vol. LXXVI(3):8-10, May-June 1988.
The Boxwood Society of the Midwest Bulletin, April 1986.
The Boxwood Bulletin, Vol. 24(1):12-13, July 1984/24(3):67, Jan. 1985.
Known Locations: Buzzards Bay Garden Club, Missouri Botanical Garden, U.S. National Arboretum, State Arboretum of Virginia.
Additional Information:
Culture and Care: Prefers dappled shade but will tolerate being sited in some direct sun.
Pests and Diseases: Resistance to leaf miner, psyllid and mites.
Available in the commercial nursery trade.

Buxus sempervirens 'Krossi-livonia'

Size (25 yrs.):	Large – 12 to 12½ feet high and 11 to 12 feet wide.
Natural Form:	Pyramidal with a slightly loose and open habit.
Annual Growth:	Fast – 6 to 6½ inches in height and 5 to 5½ inches in width.
Leaf Color:	Medium green.
Leaf Shape:	Lanceolate with some being rotund; acute and obtuse tip; cuneate base.
Leaf Size:	Medium – ⅝ to ⅞ inch long and ⅜ and 7/16 inch wide.
Leaf Surface:	Glabrous and smooth.
Internodal Length:	Long – ⅜ to ⅝ inch.
Flowering Habit:	Floriferous and heavy fruiting.
Hardiness:	Zones 6 to 8.
Plant Use:	Specimen, grouping for background and area separations.

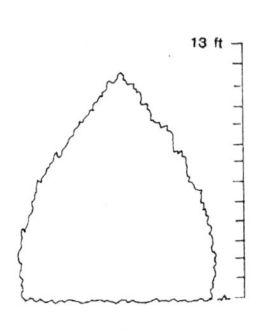

Registration: Not registered.
History: Possibly originated as a seedling of Dr. Edgar Anderson's collection from the Balkans during the early 1930s. Appeared in Kingsville Nurseries, Kingsville, Md., catalog in 1971.
Bibliography: Not documented
Known Locations: Washington Park Arboretum, U.S. National Arboretum, State Arboretum of Virginia.
Additional Information:
Culture and Care: Demonstrates no special cultural requirements.
Pests and Diseases: Indicates resistance to leaf miner, psyllid and mites in the more humid climates; no serious diseases.
Not available in the commercial nursery trade.

Buxus sempervirens 'Latifolia'

Size (25 yrs.):	Large – 6 to 6½ feet high and 7 to 7½ feet wide.
Natural Form:	Pyramidal and billowy.
Annual Growth:	Fast – 2½ to 3 inches in height and 3 to 3½ inches in width.
Leaf Color:	Dark green.
Leaf Shape:	Lanceolate and revolute; obtuse tip with some slightly retuse; cuneate base.
Leaf Size:	Medium to Large – ⅞ to 1⅛ long and ⁷⁄₁₆ to ⁹⁄₁₆ wide.
Leaf Surface:	Glabrous and smooth.
Internodal Length:	Medium – ⅜ to ½ inch.
Flowering Habit:	Moderate flowering and moderate fruiting.
Hardiness:	Zones 6 to 8.
Plant Use:	Specimen, grouping for background and area separations.

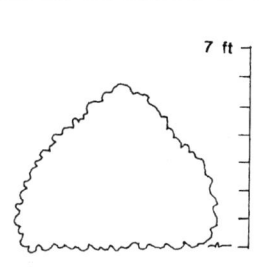

7 ft

Registration: Not registered.
History: Not documented
Bibliography: Not documented
Known Locations: Buzzards Bay Garden Club, Royal Botanic Gardens, Kew, State Arboretum of Virginia.
Additional Information:
Culture and Care: Demonstrates no special cultural requirements.
Pests and Diseases: Indicates resistance to leaf miner, psyllid and mites in the more humid climates; no serious diseases.
Not available in the commercial nursery trade.

Selected Species, Varieties, Cultivars

Buxus sempervirens 'Latifolia Marginata'

Size (25 yrs.):	Medium – 5½ to 6 feet high and 6½ to 7 feet wide.
Natural Form:	Mounded, somewhat loose and open habit.
Annual Growth:	Medium – 2½ to 3 inches in height and 3 to 3½ inches in width.
Leaf Color:	Medium green with silver variegations on the margins.
Leaf Shape:	Lanceolate to elliptic and moderately revolute; acute to obtuse tip; cuneate base.
Leaf Size:	Large – 13/16 to 1 inch long and 3/8 to ½ inch wide.
Leaf Surface:	Bullate, puckered and slightly distorted.
Internodal Length:	Long – 3/8 to 5/8 inch.
Flowering Habit:	Not observed.
Hardiness:	Zones 6 to 8.
Plant Use:	Specimen.

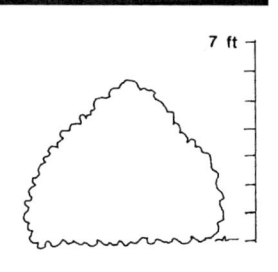

Registration: *Kew Handlist of Trees and Shrubs*, 269, 1925.
History: Most likely originated as an open-pollinated seedling or mutation somewhere in Europe.
Bibliography:
The Boxwood Bulletin, Vol. 9(1):7, July 1969.
Known Locations: Washington Park Arboretum, U.S. National Arboretum, State Arboretum of Virginia.
Additional Information:
Culture and Care: Demonstrates no special cultural requirements.
Pests and Diseases: Indicates resistance to leaf miner, psyllid and mites in the more humid climates; no serious diseases.
Propagation: Somewhat difficult and quite slow to root. Best to take cuttings when spring growth has hardened and include a portion of the heel from the previous year's growth; bruise the stem tissue or slit each side of the stem; use an IBA powder dip.
Not available in the commercial nursery trade.

Buxus sempervirens 'Liberty'

Size (25 yrs.):	Medium – 4½ to 5 feet in height and 3½ to 4 feet in width.
Natural Form:	Ovate.
Annual Growth:	Medium – 1½ to 2 inches in height and 1 to 1½ inches in width.
Leaf Color:	Dark green.
Leaf Shape:	Lanceolate – upright habit; obtuse tip, some retuse; cuneate base.
Leaf Size:	Large – ¾ to 1 inch long and 5/16 to ½ inch wide.
Leaf Surface:	Glabrous – smooth, tending toward matte.
Internodal Length:	Long – ⅜ to ½ inch.
Flowering Habit:	Moderate flowering and fruiting.
Hardiness:	Zones 5 to 8.
Plant Use:	Specimen, foundation planting, grouping for background and area separations, hedgings.

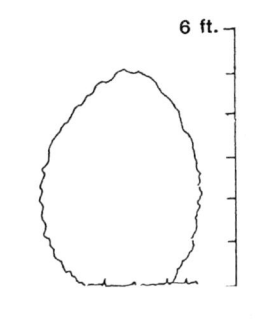

6 ft.

Registration: Not registered.
History: Original cuttings were brought from Michigan by Mrs. Schrader of Schrader Peony Gardens, Liberty, Ind., to ultimately border the peony beds. James A. Clark, Indiana state nursery inspector, observed these boxwood plants for several years and particularly noted their hardiness after a severe winter in 1955. In 1956, Mr. Clark propagated some 25 cuttings from these plants and by 1961 he named the plant 'Liberty' and commenced producing additional plants released by Cunningham Gardens, Waldron, Ind.
Bibliography: Not documented
Known Locations: State Arboretum of Virginia.
Additional Information:
Culture and Care: Prefers dappled shade, will tolerate being sited in some direct sun.
    Breaks dormancy late.
Pests and Diseases: Resistance to leaf miner, psyllid and mites.
Available in the commercial nursery trade.

Buxus sempervirens 'Macrophylla'

Size (25 yrs.):	Large – 7 to 7½ feet high and 7½ to 8 feet wide.
Natural Form:	Pyramidal with a very loose and open habit.
Annual Growth:	Fast – 3½ to 4 inches in height and 3½ to 4 inches in width.
Leaf Color:	Dark green.
Leaf Shape:	Generally ovate and revolute; acute to obtuse tip; cuneate base.
Leaf Size:	Medium – ¾ to ¹³/₁₆ inch long and ½ to ¾ inch wide.
Leaf Surface:	Glossy and smooth.
Internodal Length:	Medium – ⁵/₁₆ to ⅜ inch.
Flowering Habit:	Floriferous with heavy fruiting.
Hardiness:	Zones 6 to 8.
Plant Use:	Specimen.

Registration: *Kew Handlist of Trees and Shrubs,* 609, 1902.
History: Most likely originated as an open-pollinated seedling somewhere in Europe.
Bibliography:
Dallimore, W. *Holly, Yew and Box,* 1908.
The Boxwood Bulletin, Vol. 2(3):37, Jan. 1963/9(1):7, July 1969.
Known Locations: Hillier's Arboretum, Royal Botanic Gardens, Kew; U.S. National Arboretum, State Arboretum of Virginia.
Additional Information:
Culture and Care: Transplants readily; prefers dappled shade but will tolerate being sited in some direct sun but occasionally suffers from winter bronzing. Demonstrates no special cultural requirements.
Pests and Diseases: Indicates resistance to leaf miner, psyllid and mites in the more humid climates; no serious diseases.
Propagation: Cuttings root quite readily without the use of an IBA powder dip; however, they root slightly faster with the dip. The poly-tent procedure usually produces rooted cuttings nearly as fast as the mist systems, about 6 to 8 weeks.
Available in the commercial nursery trade.

Buxus sempervirens 'Maculata'

Size (25 yrs.):	Large – 7 to 8 feet high and 7 to 8 feet wide.
Natural Form:	Pyramidal with a very loose and open habit.
Annual Growth:	Fast – $3^1/_2$ to 4 inches in height and $3^1/_2$ to 4 inches in width.
Leaf Color:	Dark green with occasional blotches of gold that decrease with age.
Leaf Shape:	Generally ovate and revolute; acute to obtuse tip; cuneate base.
Leaf Size:	Medium – $^3/_4$ to $^{13}/_{16}$ inch long and $^1/_2$ to $^3/_4$ inch wide.
Leaf Surface:	Glossy and smooth.
Internodal Length:	Medium – $^5/_{16}$ to $^3/_8$ inch.
Flowering Habit:	Floriferous with heavy fruiting.
Hardiness:	Zones 6 to 8.
Plant Use:	Specimen, foundation planting, grouping for background and area separations, hedgings.

Registration: *Kew Handlist of Trees and Shrubs Grown in Arboretum,* Part II:131, 1896. Invalid names: 'Latifolia Aurea Maculata,' 'Japonica Aurea,' 'Japonica Aurea Rotundifolia.'
History: Most likely originated as an open-pollinated seedling somewhere in Europe. Often found in England's churchyards.
Bibliography: Not documented
Known Locations: State Arboretum of Virginia.
Additional Information:
Culture and Care: Demonstrates no special cultural requirements.
Pests and Diseases: Indicates resistance to leaf miner, psyllid and mites in the more humid climates; no serious diseases.
Available in the commercial nursery trade.

Buxus sempervirens 'Mary Gamble'

Size (25 yrs.):	Small – 3 to 4 feet high and 3 to 4 feet wide.
Natural Form:	Spherical.
Annual Growth:	Slow – 1 to 1½ inches in height and 1 to 1½ inches in width.
Leaf Color:	Medium green.
Leaf Shape:	Lanceolate – slightly revolute; obtuse tip; cuneate base.
Leaf Size:	Small – ½ to ¹¹/₁₆ inch long and ³/₁₆ to ⁵/₁₆ inch wide.
Leaf Surface:	Glabrous – smooth.
Internodal Length:	Short – ¼ to ⁵/₁₆ inch.
Flowering Habit:	Not observed.
Hardiness:	Zones 5 to 8.
Plant Use:	Specimen, grouping for background and area separations, edgings, hedgings.

5 ft.

Registration: The Boxwood Society of the Midwest in *The Boxwood Bulletin,* Vol. 26(2):34-35, Oct. 1986.
History: Parent plant may well have had its beginning in Westfield, Mass., in the early 1800s, moved on to Weston, Mass., and on to Lincoln, Mass., where it was discovered by Lucy Mason of St. Louis, Mo., sited in the garden of Peggy Marsh. In 1971 Lucy Mason brought cuttings back to be tested by The Boxwood Society of the Midwest. The cultivar was named in 1986 for Mary A. Gamble, first president and one of the founders of The Boxwood Society of the Midwest.
Bibliography:
Missouri Botanical Garden Bulletin, Vol. LXXVI(3):8-10, May-June 1988.
The Boxwood Society of the Midwest Bulletin, April 1986.
The Boxwood Bulletin, Vol. 26(2):34-35, Oct. 1986/27(2):36, Oct. 1987.
Known Locations: Missouri Botanical Garden, State Arboretum of Virginia.
Additional Information:
Culture and Care: Prefers dappled shade but will tolerate siting in some direct sun.
    Breaks dormancy late.
Pests and Diseases: Resistance to leaf miner, pysllid and mites.
Available in the commercial nursery trade.

Buxus sempervirens 'Memorial'

Size (25 yrs.):	Medium – 4 to 5 feet high and 2 to 2½ feet wide. The mother plant, about 60 years of age, measured 8 feet high and 9 feet wide.
Natural Form:	Somewhat columnar tending toward ovate in its early years, tends to become more pyramidal and billowy after 35-40 years.
Annual Growth:	Medium – 2 to 2½ inches in height and 1 to 1½ inches in width.
Leaf Color:	Medium green.
Leaf Shape:	Lanceolate – slightly revolute; retuse tip; cuneate base.
Leaf Size:	Large – ⅞ to 1 inch long and 3/16 to ½ inch wide.
Leaf Surface:	Glabrous – smooth.
Internodal Length:	Medium – ¼ to ⅜ inch.
Flowering Habit:	Not observed.
Hardiness:	Zones 6 to 8.
Plant Use:	Specimen, grouping for background and area separations, hedgings.

Registration: Dr. J.T. Baldwin, Jr. in *The Boxwood Bulletin*, Vol. 6(4):ibc, April 1967.
History: Mother plant was located at Cedar Grove Cemetery, Williamsburg, Va. Named by Merlin C. Larson of Williamsburg.
Bibliography:
The Boxwood Bulletin, Vol. 6(4):ibc, April 1967/15(4):55-58, April 1976/19(2):26, Oct. 1979/29(3):41-42, Jan. 1989.
Known Locations: College of William and Mary, U.S. National Arboretum, State Arboretum of Virginia.
Additional Information:
Culture and Care: Prefers dappled shade but will tolerate siting in some direct sun.
Pests and Diseases: Resistance to leaf miner, psyllid and mites.
Available in the commercial nursery trade.

Buxus sempervirens 'Myosotidifolia'

Size (25 yrs.):	Large – 8½ to 9 feet high and 9½ to 10 feet wide.
Natural Form:	Pyramidal, billowy with upright growth habit and moderately compact.
Annual Growth:	Fast – 4 to 4½ inches in height and 4½ to 5 inches in width.
Leaf Color:	Medium green.
Leaf Shape:	Lanceolate and slightly revolute; acute tip; cuneate base.
Leaf Size:	Large – 1⅛ to 1¼ inches long and 5/16 to ⅝ inch wide.
Leaf Surface:	Glabrous and smooth.
Internodal Length:	Long – 7/16 to ⅝ inch.
Flowering Habit:	Floriferous and heavy fruiting.
Hardiness:	Zones 6 to 8.
Plant Use:	Specimen, grouping for background and area separations.

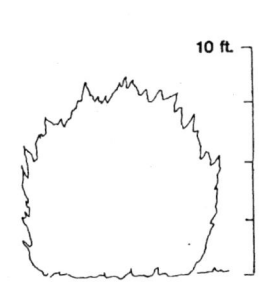

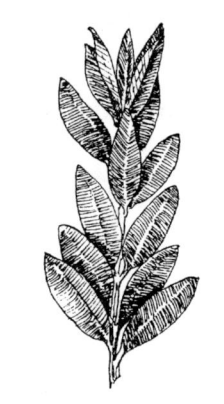

Registration: *Kew Handlist of Trees and Shrubs Grown in Arboretum*, Part II:131, 1896.
History: Most likely originated as an open-pollinated seedling somewhere in Europe.
Bibliography:
Bailey, L.H. *Hortus Third*, 1976.
Dallimore, W. *Holly, Yew and Box*, 1908.
Krussman, G. *Manual of Cultivated Broad-leaved Trees and Shrubs*, Vol. I, A-D, 1984.
The Boxwood Bulletin, Vol. 2(3):37, Jan. 1963/9(1):8, July 1969.
Known Locations: Hillier's Arboretum, Washington Park Arboretum, U.S. National Arboretum, State Arboretum of Virginia.
Additional Information:
Culture and Care: Demonstrates no special cultural requirements.
Pests and Diseases: Indicates resistance to leaf miner, psyllid and mites in the more humid climates; no serious diseases.
Available in the commercial nursery trade.

Buxus sempervirens 'Myrtifolia'

Size (25 yrs.):	Medium – 3 to 4 feet high and 2^1/$_2$ to 3 feet wide.
Natural Form:	Pyramidal – tending toward ovate, compact habit.
Annual Growth:	Medium – 1^1/$_2$ to 2 inches in height and 1 to 1^1/$_2$ inches in width.
Leaf Color:	Medium green.
Leaf Shape:	Elliptic – slightly revolute; acute tip; cuneate base.
Leaf Size:	Medium – ^{1}/$_2$ to ^{13}/$_{16}$ inch long and ^{1}/$_4$ to ^{5}/$_{16}$ inch wide.
Leaf Surface:	Glossy – smooth.
Internodal Length:	Medium – ^{5}/$_{16}$ to ^{3}/$_8$ inch.
Flowering Habit:	Not observed.
Hardiness:	Zones 5(protected) to 8.
Plant Use:	Specimen, foundation planting, grouping for background and area separations, hedgings.

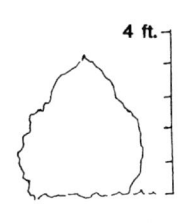

Registration: *Catalog of Trees, Shrubs, Plants, Flower Roots, Seeds, & c.,* Gordon, Dermer and Edmonds Pl.6.1782. Invalid names: 'Myrtifolia Suffruticosa' and 'Leptophylla.'
History: Most likely originated as an open-pollinated seedling somewhere in Europe.
Bibliography:
Bailey, L.H. *Hortus Third,* 1976.
Dirr, M.A. *Manual of Woody Landscape Plants,* 4th Ed., 1990.
Everett, T.H. *The New York Botanical Garden Ill. Encyc. of Hort.,* Vol. 2, 1981.
Hotte, A.C. *The Book of Shrubs,* 1931.
Krussman, G. *Manual of Cultivated Broad-leaved Trees and Shrubs,* Vol. I, A-D, 1984.
Wright, M. *The Complete Handbook of Garden Plants,* 1984.
Missouri Botanical Garden Bulletin, Vol. LXXVI(3):8-10, May-June 1988.
The Boxwood Bulletin, Vol. 2(3):37, Jan. 1963/3(2):24, Oct. 1963/9(1):8, July 1969/ 14(1):12, July 1974/17(2):26, Oct. 1977/20(3):42,45, Jan. 1981.
Known Locations: Hillier's Arboretum, Royal Botanic Gardens, Kew; Washington Park Arboretum, U.S. National Arboretum, State Arboretum of Virginia.
Additional Information:
Culture and Care: An excellent plant for city lots. Tolerates siting in direct sun quite well, but protected from winter winds. Breaks dormany late.
Pests and Diseases: Resistance to leaf miner, psyllid and mites.
Available in the commercial nursery trade.

Selected Species, Varieties, Cultivars

Buxus sempervirens 'Natchez'

Size (25 yrs.):	Dwarf – 2 to 2½ feet high and 2½ to 3 feet wide.
Natural Form:	Mounded.
Annual Growth:	Slow – 1¼ to 1½ inches in height and 1½ to 1¾ inches in width.
Leaf Color:	Medium green with bluish hue.
Leaf Shape:	Elliptic to obovate; acute to obtuse tip; cuneate base.
Leaf Size:	Medium – ⅝ to ¾ inch long and ⅜ to ½ inch wide.
Leaf Surface:	Glabrous and smooth.
Internodal Length:	Medium – ¼ to 5/16 inch.
Flowering Habit:	Not observed.
Hardiness:	Zones 5 to 8.
Plant Use:	Specimen, grouping for background and area separations, edgings, hedgings.

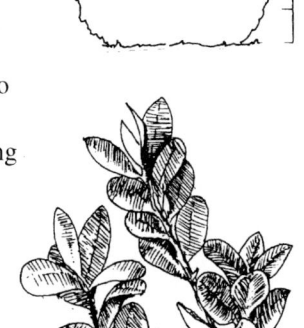

Registration: Mary A. Gamble in *The Boxwood Bulletin*, Vol. 26(3):62-63, Jan. 1987.
History: Came from the nursery of Clarence Barbré, Webster Grove, Mo. The plant's origin is unknown.
Bibliography:
The Boxwood Bulletin, Vol. 26(3):62-63, Jan. 1987.
Known Locations: Missouri Botanical Garden, State Arboretum of Virginia.
Additional Information:
Culture and Care: Demonstrates no special cultural requirements.
Pests and Diseases: Indicates resistance to leaf miner, psyllid and mites in the more humid climates; no serious diseases.
Available in the commercial nursery trade.

Buxus sempervirens 'Newport Blue'

Size (25 yrs.):	Medium – 4 to 5 feet high and 5 to 6 feet wide.
Natural Form:	Somewhat pyramidal, loose and open habit.
Annual Growth:	Medium – $1^1/_2$ to 2 inches in height and $2^1/_2$ to 3 inches in width.
Leaf Color:	Medium green.
Leaf Shape:	Lanceolate; acute tip; cuneate base.
Leaf Size:	Medium, tending toward large $^7/_8$ to 1 inch long and $^5/_{16}$ to $^3/_8$ inch wide.
Leaf Surface:	Glossy and shiny.
Internodal Length:	Medium – $^3/_8$ to $^1/_2$ inch.
Flowering Habit:	Not observed.
Hardiness:	Zones 6 to 8.
Plant Use:	Specimen, grouping for background and area separations.

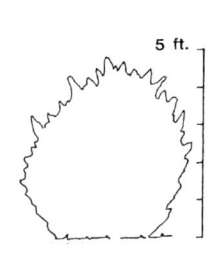

Registration: Catalog, Boulevard Nurseries, Newport, R.I., 1941. Two forms listed – globe and pyramidal.
History: Believed to be an open-pollinated seedling selected by Boulevard Nurseries in the 1930s.
Bibliography:
Bush-Brown, J. and L. *America's Garden Book*, Rev., New York Botanical Garden, 1980.
Dirr, M.A. *Manual of Woody Landscape Plants*, 4th Ed., 1990.
Wyman, D. *Wyman's Gardening Encyclopedia*, New Exp. Ed., 1986.
Known Locations: Buzzards Bay Garden Club, U.S. National Arboretum, State Arboretum of Virginia.
Additional Information:
Culture and Care: Demonstrates no special cultural requirements.
Pests and Diseases: Indicates resistance to leaf miner, psyllid and mites in the more humid climates; no serious diseases.
Available in the commercial nursery trade.

Selected Species, Varieties, Cultivars

Buxus sempervirens 'Nish'

Size (25 yrs.):	Medium – 3½ to 4 feet high and 4 to 4½ feet wide.
Natural Form:	Pyramidal, moderately billowy, somewhat open habit.
Annual Growth:	Medium – 1½ to 2 inches in height and 1½ to 2 inches in width.
Leaf Color:	Dark green with yellow undertone.
Leaf Shape:	Elliptical to lanceolate and slightly revolute; acute tip; cuneate base.
Leaf Size:	Medium – ⁵/₈ to ⁷/₈ inch long and ⁵/₁₆ to ³/₈ inch wide.
Leaf Surface:	Glabrous and smooth.
Internodal Length:	Medium – ¼ to ³/₈ inch.
Flowering Habit:	Not observed.
Hardiness:	Zones 5 to 8.
Plant Use:	Specimen, foundation planting, grouping for background and area separations, hedgings.

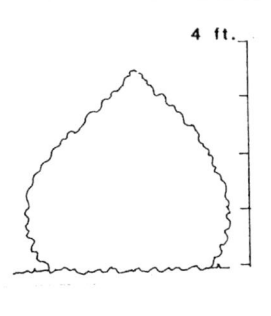

Registration: Mary A. Gamble in *The Boxwood Bulletin,* 14(4):61, April 1975.
History: Originated from seed sent to Dr. Edgar Anderson from Yugoslavia in 1935. It is named after a town in Yugoslavia, now written Nic.
Bibliography:
Missouri Botanical Garden Bulletin, Vol. LXXVI(3):8-10, May-June 1988.
The Boxwood Bulletin, Vol. 14(4):cover, April 1975/24(2):50, Oct. 1984.
Known Locations: Longwood Gardens, Missouri Botanical Garden, U.S. National Arboretum, State Arboretum of Virginia.
Additional Information:
Culture and Care: Demonstrates no special cultural requirements.
Pests and Diseases: Indicates resistance to leaf miner, psyllid and mites in the more humid climates; no serious diseases.
Not available in the commercial nursery trade.

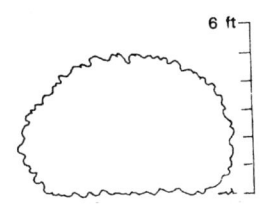

Buxus sempervirens 'Northern Find'

Size (25 yrs.):	Large – 5 to 6 feet high and 8 to 9 feet wide.
Natural Form:	Mounded, somewhat open habit.
Annual Growth:	Fast – 2½ to 3 inches in height and 3½ to 4 inches in width.
Leaf Color:	Dark green with an occasional silver variegation on the leaf margin.
Leaf Shape:	Lanceolate and slightly revolute; acute to obtuse tip; cuneate base.
Leaf Size:	Large – 1 to 1¼ inches long and 7/16 to 9/16 inch wide.
Leaf Surface:	Glabrous and smooth.
Internodal Length:	Long – 3/8 to 7/16 inch.
Flowering Habit:	Moderate flowering and fruit set.
Hardiness:	Zones 5 to 8.
Plant Use:	Specimen, foundation planting, grouping for background and area separations, hedgings.

Registration: D. Wyman in *Arnoldia*, 23(5):87-88, 1963.
History: The original clone was discovered growing on the grounds of St. Joseph's Hospital in Hamilton, Ontario in the mid-1930s. It was named and released commercially by Woodland Nurseries Ltd. of Cooksville, Ontario, Canada in 1955.
Bibliography:
Dirr, M.A. *Manual of Woody Landscape Plants*, 4th Ed., 1990.
Wyman, D. *Wyman's Gardening Encyclopedia*, New Expanded 2nd Ed., 1986.
The Boxwood Bulletin, Vol. 3(1):1, July 1963/5(1):1, July 1965.
Known Locations: Arnold Arboretum, U.S. National Arboretum, State Arboretum of Virginia.
Additional Information:
Culture and Care: Demonstrates no special cultural requirements.
Pests and Diseases: Indicates resistance to leaf miner, psyllid and mites in the more humid climates; no serious diseases.
Not available in the commercial nursery trade.

Selected Species, Varieties, Cultivars

Buxus sempervirens 'Northland'

Size (25 yrs.):	Large – 6 to 6½ feet high and 7 to 7½ feet wide.
Natural Form:	Pyramidal – billowy compact habit.
Annual Growth:	Medium – 2½ to 3 inches in height and 3 to 3½ inches in width.
Leaf Color:	Dark green.
Leaf Shape:	Lanceolate to elliptic – slightly revolute; acute to obtuse tip; cuneate base.
Leaf Size:	Medium – ¾ to 1 1/16 inches long and 5/16 to 3/8 inch wide.
Leaf Surface:	Glabrous – smooth.
Internodal Length:	Medium – ¼ to 5/16 inch.
Flowering Habit:	Floriferous and heavy fruiting.
Hardiness:	Zones 5 to 8.
Plant Use:	Specimen, foundation planting, grouping for background and area separations, hedgings.

7 ft

Registration: Catalog, C.W. Stuart and Company, Newark, N.Y., 1949.
History: Discovered as an open-pollinated seedling by the C.W. Stewart Company of Newark, N.Y., in the early 1930s. Named and commercially released in 1949.
Bibliography:
Dirr, M.A. *Manual of Woody Landscape Plants*, 4th Ed., 1990.
Flint, H.L. *Horticulture,* March 1987.
Wyman, D. *Wyman's Gardening Encyclopedia*, New Expanded 2nd Ed., 1986.
The Boxwood Bulletin, Vol. 21(3):45, Jan. 1982.
Known Locations: Buzzards Bay Garden Club, U.S. National Arboretum, State Arboretum of Virginia.
Additional Information:
Culture and Care: Will tolerate siting in some direct sun. Breaks dormancy late.
Pests and Diseases: Resistance to leaf miner, psyllid and mites.
Available in the commercial nursery trade.

Buxus sempervirens 'Notata'

Size (25 yrs.):	Large – 8 to 9 feet high and 11 to 12 feet wide.
Natural Form:	Somewhat mounded and not one of the better cultivars; does not appear to be particularly stable.
Annual Growth:	Fast – 4 to 5 inches in height and 5 to 6 inches in width.
Leaf Color:	Dark green and bi-colored gold at the apex, especially on older specimens.
Leaf Shape:	Generally lanceolate and revolute; somewhat obtuse tip; cuneate base.
Leaf Size:	Medium – $3/4$ to $13/16$ inch long and $1/4$ to $1/2$ inch wide.
Leaf Surface:	Glabrous and smooth.
Internodal Length:	Medium – $1/2$ to $13/16$ inch.
Flowering Habit:	Not observed.
Hardiness:	Zones 6 to 8.
Plant Use:	Specimen.

Registration: R. Weston in *Botanicus Universalis*, 1:31, 1770. Invalid name 'Gold Tip.'
History: Not documented
Bibliography:
Krussman, G. *Manual of Cultivated Broad-leaved Trees and Shrubs*, Vol. I, A-D, 1986.
The Boxwood Bulletin, Vol. 9(1):8, July 1969.
Known Locations: U.S. National Arboretum, State Arboretum of Virginia.
Additional Information:
Culture and Care: Demonstrates no special cultural requirements.
Pests and Diseases: Indicates resistance to leaf miner, psyllid and mites in the more humid climates; no serious diseases.
Not available in the commercial nursery trade.

Buxus sempervirens 'Pendula'

Size (25 yrs.):	Large – 8 to 9 feet high and 6 to 7 feet wide.
Natural Form:	Unusual – tree-like with a dominant leader and pendulous branches.
Annual Growth:	Fast – $3^1/_2$ to $4^1/_2$ inches in height and 3 to $3^1/_2$ inches in width.
Leaf Color:	Medium green with yellowish undertone.
Leaf Shape:	Broadly lanceolate and slightly revolute; acute tip; cuneate base.
Leaf Size:	Long – 1 to $1^1/_8$ inches long and $^5/_{16}$ to $^3/_8$ inch wide.
Leaf Surface:	Glabrous and smooth.
Internodal Length:	Long – $^3/_4$ to $^7/_8$ inch.
Flowering Habit:	Floriferous and heavy fruiting.
Hardiness:	Zones 5(protected) to 8.
Plant Use:	Specimen.

Registration: Catalog, Simon Louis 21, 1869. Invalid name 'Aurea Maculata Pendula.'
History: Most likely originated as an open-pollinated seedling somewhere in Europe. Sometimes referred to as weeping boxwood.
Bibliography:
Bailey, L.H. *Hortus Third*, 1976.
Dallimore, W. *Holly, Yew and Box*, 1908.
Dirr, M.A. *Manual of Woody Landscape Plants*, 4th Ed., 1990.
Hotte, A.C. *The Book of Shrubs*, 1931.
Krussman, G. *Manual of Cultivated Broad-leaved Trees and Shrubs*, Vol. I, A-D, 1984.
Wyman, D. *Wyman's Gardening Encyclopedia*, New Expanded 2nd Ed., 1986.
The Boxwood Bulletin, Vol. 2(3):37, Jan. 1963/4(2):26, Oct. 1964/9(1):8, July 1969/ 16(1):15, July 1976/20(3):48, Jan. 1981/21(4):65, April 1982/27(2):38, Oct. 1987.
Known Locations: Arnold Arboretum, Buzzards Bay Garden Club, College of William and Mary, Hillier's Arboretum, Royal Botanic Gardens, Kew; Longwood Gardens, Secrest Arboretum, Washington Park Arboretum, U.S. National Arboretum, State Arboretum of Virginia.
Additional Information:
Culture and Care: Demonstrates no special cultural requirements.
Pests and Diseases: Indicates resistance to psyllid; no serious diseases.
Available in the commercial nursery trade.

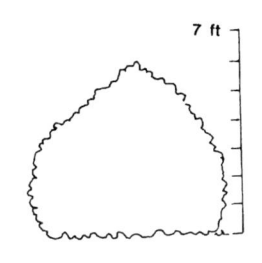

Buxus sempervirens 'Ponteyi'

Size (25 yrs.):	Large – 6 to 7 feet high and 7 to 8 feet wide. A 30-year-old specimen measures 8 feet high and 9 feet wide.
Natural Form:	Pyramidal, billowy and tending toward mounded, compact habit.
Annual Growth:	Medium – 3 to $3^1/_2$ inches in height and $3^1/_2$ to 4 inches in width.
Leaf Color:	Dark green.
Leaf Shape:	Lanceolate and moderately revolute; acute tip tending toward obtuse; cuneate base.
Leaf Size:	Large – 1 to $1^3/_{16}$ inches long and $^3/_8$ to $^5/_8$ inch wide.
Leaf Surface:	Glabrous and smooth, occasionally has a few variegated branchlets.
Internodal Length:	Long – $^7/_{16}$ to $^5/_8$ inch.
Flowering Habit:	Floriferous and heavy fruiting.
Hardiness:	Zones 6 to 8.
Plant Use:	Specimen, foundation planting, grouping for background and area separations, hedgings, topiary.

Registration: L. Dippel in *Handbuch der Laubholzkunde* 3:81, 1893.
History: Most likely originated as an open-pollinated seedling somewhere in Europe.
Bibliography: Not documented
Known Locations: Royal Botanic Gardens, Kew; U.S. National Arboretum, State Arboretum of Virginia.
Additional Information:
Culture and Care: Demonstrates no special cultural requirements.
Pests and Diseases: Indicates resistance to leaf miner, psyllid and mites in the more humid climates; no serious diseases.
Available in the commercial nursery trade.

Buxus sempervirens 'Prostrata'

Size (25 yrs.):	Medium – 5½ to 6 feet high and 6½ to 7 feet wide. A 30-year-old specimen measures 7 feet high and 7½ feet wide.
Natural Form:	Unusual – loose and open habit with tendency to droop.
Annual Growth:	Medium – 2½ to 3 inches in height and 2¾ to 3¼ inches in width.
Leaf Color:	Dark green.
Leaf Shape:	Elliptic to lanceolate and slightly revolute; acute tip tending toward obtuse; cuneate base.
Leaf Size:	Large – ⅞ to 1 inch long and ⅜ to 7/16 inch wide.
Leaf Surface:	Glabrous and occasionally puckered.
Internodal Length:	Long – ⅜ to 7/16 inch.
Flowering Habit:	Floriferous and heavy fruiting.
Hardiness:	Zones 6 to 8.
Plant Use:	Specimen.

Registration: W. Bean in *Trees and Shrubs Hardy in the British Isles*, 1:278, 1914. Invalid name 'Horizontalis.'
History: Most likely originated as an open-pollinated seedling somewhere in Europe.
Bibliography:
Bailey, L.H. *Hortus Third*, 1976.
Dallimore, W. *Holly, Yew and Box*, 1908.
The Boxwood Bulletin, Vol. 2(3):37, Jan. 1963/9(1):8, July 1969/27(3):65, Jan. 1988.
Known Locations: Royal Botanic Gardens, Edinburgh; Hillier's Arboretum, Royal Botanic Gardens, Kew; Washington Park Arboretum, U.S. National Arboretum, State Arboretum of Virginia.
Additional Information:
Culture and Care: Demonstrates no special cultural requirements.
Pests and Diseases: Indicates resistance to leaf miner, psyllid and mites in the more humid climates; no serious diseases.
Not available in the commercial nursery trade.

Buxus sempervirens 'Pullman'

Size (25 yrs.):	Large – 6 to 7 feet high and 6 to 7 feet wide.
Natural Form:	Somewhat conical with open habit.
Annual Growth:	Medium – 2½ to 3 inches in height and 2½ to 3 inches in width.
Leaf Color:	Dark green with black hue.
Leaf Shape:	Elliptic to lanceolate – slightly revolute; acute tip; cuneate base.
Leaf Size:	Medium – 7/16 to 7/8 inch long and 5/16 to 3/8 inch wide.
Leaf Surface:	Glabrous – smooth.
Internodal Length:	Medium – ¼ to 5/16 inch.
Flowering Habit:	Not observed.
Hardiness:	Zones 5 to 8.
Plant Use:	Specimen, foundation planting, grouping for background and area separations, hedgings.

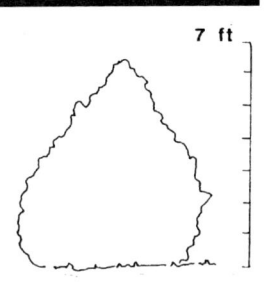

Registration: William Pullman in *The Boxwood Bulletin*, Vol. 11(2):20-21, Oct. 1971.
History: Discovered in the garden of William A.P. Pullman in Lake Forest, Ill., about 1955 as an open-pollinated seedling.
Bibliography:
M.A. Dirr. *Manual of Woody Landscape Plants*, 4th Ed., 1990.
H.L. Flint. *Horticulture*, March 1987.
The Boxwood Society of the Midwest Bulletin, April 1986.
Missouri Botanical Garden Bulletin, Vol. LXXVI(3):8-10, May-June 1988.
The Boxwood Bulletin, Vol. 11(2):20-21, Oct. 1971/16(4):57, April 1977/20(3):42,44, Jan. 1981/23(1):18-19, July 1983.
Known Locations: Arnold Arboretum, Buzzards Bay Garden Club, Missouri Botanical Garden, State Arboretum of Virginia.
Additional Information:
Culture and Care: Prefers dappled shade but will tolerate siting in some direct sun.
   Breaks dormancy late.
Pests and Diseases: Indicates resistance to leaf miner, prone to minor psyllid infestation; no serious diseases.
Available in the commercial nursery trade.

Buxus sempervirens 'Pyramidalis'

Size (25 yrs.):	Large – 8 to 9 feet high and 4 to 4½ feet wide.
Natural Form:	Conical – loose open habit.
Annual Growth:	Fast – 4½ to 5 inches in height and 2 to 2½ inches in width.
Leaf Color:	Medium green.
Leaf Shape:	Elliptic – acute tip tending toward obtuse, somewhat revolute; cuneate base.
Leaf Size:	Large – 1 to 1⅛ inches long and ⅜ to ½ inch wide.
Leaf Surface:	Glabrous – smooth.
Internodal Length:	Medium – ¼ to ⅜ inch.
Flowering Habit:	Floriferous with heavy fruiting.
Hardiness:	Zones 6 to 8.
Plant Use:	Specimen, grouping for background and area separations, allees.

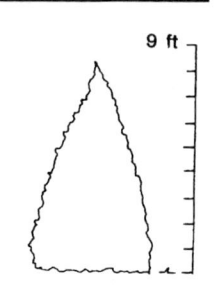

Registration: Catalog, Simon Louis, 21, 1869. Invalid names: 'Oleaefolia Elegans,' 'Pyramidata.'
History: Most likely originated as an open-pollinated seedling somewhere in Europe.
Bibliography:
Bailey, L.H. *Hortus Third*, 1976.
Dallimore, W. *Holly, Yew and Box*, 1908.
Everett, T.H. *The New York Botanical Garden Ill. Encyc. of Hort.*, Vol. 2, 1981.
Hotte, A.C. *The Book of Shrubs*, 1931.
Krussman, G. *Manual of Cultivated Broad-leaved Trees and Shrubs*, Vol. I, A-D, 1984.
The Boxwood Bulletin, Vol. 2(3):37, Jan. 1963/9(1):8, July 1969/27(3):65, Jan. 1988/ 27(4):cover, April 1988.
Known Locations: Arnold Arboretum, Longwood Gardens, Washington Park Arboretum, U.S. National Arboretum, State Arboretum of Virginia.
Additional Information:
Culture and Care: Tolerates siting in direct sun quite well. Breaks dormancy late.
Pests and Diseases: Prone to leaf miner and psyllid.
Available in the commercial nursery trade.

Buxus sempervirens 'Pyramidalis Hardwickensis'

Size (25 yrs.):	Large – 10 to 12 feet high and 3 to 4 feet wide. A 40-year-old specimen measures 15 feet high and 7 feet wide.
Natural Form:	Conical, stiffly upright, loose and open habit.
Annual Growth:	Fast – 5 to 5$^1/_2$ inches in height and 1$^1/_2$ to 2 inches in width.
Leaf Color:	Dark green.
Leaf Shape:	Obovate to elliptic and slightly revolute; acute tip; cuneate base.
Leaf Size:	Large – $^{13}/_{16}$ to 1$^1/_8$ inches long and $^3/_8$ to $^5/_8$ inch wide.
Leaf Surface:	Glabrous and smooth.
Internodal Length:	Long – $^5/_8$ to $^7/_8$ inch.
Flowering Habit:	Not observed.
Hardiness:	Zones 5 to 8.
Plant Use:	Specimen, grouping for background and area separations, allees.

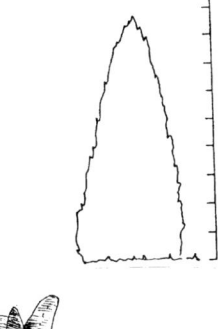

Registration: *Kew Handlist of Trees and Shrubs,* 269, 1925.
History: Most likely originated as an open-pollinated seedling somewhere in Europe.
Bibliography:
Dallimore, W. *Holly, Yew and Box,* 1908.
The Boxwood Society of the Midwest Bulletin, April 1986.
The Boxwood Bulletin, Vol. 4(4):64-66, April 1965/9(4): 53, April 1970/20(3):43,45, Jan. 1981.
Known Locations: College of William and Mary, Missouri Botanical Garden.
Additional Information:
Culture and Care: Demonstrates no special cultural requirements.
Pests and Diseases: Indicates resistance to leaf miner, psyllid and mites in the more humid climates; no serious diseases.
Not available in the commercial nursery trade.

Selected Species, Varieties, Cultivars

Buxus sempervirens 'Rochester'

Size (25 yrs.):	Medium – 3 to 4 feet high and 4 feet wide.
Natural Form:	Pyramidal.
Annual Growth:	Slow – up to 1½ inches in height and 2 inches in width.
Leaf Color:	Medium green.
Leaf Shape:	Elliptic; acute tip; cuneate base.
Leaf Size:	Medium – ¾ to ⅞ inch long and ¼ to ½ inch wide.
Leaf Surface:	Glabrous and smooth.
Internodal Length:	Medium – ¼ to ⅜ inch.
Flowering Habit:	Not observed.
Hardiness:	Zones 5 to 8.
Plant Use:	Specimen, foundation planting, grouping for background and area separations, hedgings.

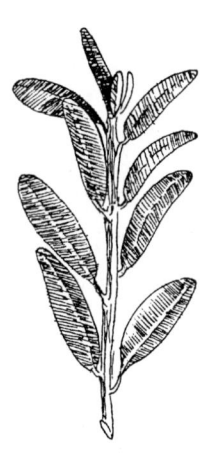

Registration: Not registered.
History: Believed to have been originally selected by the Monroe County Department of Parks, Rochester, N.Y., in the late 1950s and first released by Girard Nurseries, Geneva, Ohio, in the late 1960s. Girard Nurseries listed a plant called 'Pride of Rochester' in its 1976 catalog. "Has a leaf like the regular *sempervirens* but a little darker and compact growing. The most important thing is it's extremely hardy which rivals the Korean boxwood." In 1988 the Girard Nurseries catalog described 'Pride of Rochester' as "Aristocratic, slow-growing compact evergreen. Good for small hedgings, urns, porch boxes and edgings. Our own hardy strain."
Bibliography: Not documented
Known Locations: State Arboretum of Virginia.
Additional Information:
Culture and Care: Demonstrates no special cultural requirements.
Pests and Diseases: Indicates resistance to leaf miner, psyllid and mites in the more humid climates; no serious diseases.
Available in the commercial nursery trade.

Buxus sempervirens 'Rotundifolia'

Size (25 yrs.):	Large – 8 to 8½ feet high and 10 to 11 feet wide.
Natural Form:	Mounded, somewhat loose and open habit.
Annual Growth:	Fast – 3 to 4 inches in height and 5 to 5½ inches in width.
Leaf Color:	Medium green.
Leaf Shape:	Rotund and slightly revolute; obtuse and retuse tip; cuneate base.
Leaf Size:	Large – ¹¹/₁₆ to 1 inch long and ⁷/₁₆ to 1 inch wide.
Leaf Surface:	Glossy and smooth.
Internodal Length:	Long – ³/₈ to ³/₄ inch.
Flowering Habit:	Floriferous and heavy fruiting.
Hardiness:	Zones 6 to 8.
Plant Use:	Specimen, foundation planting, grouping for background and area separations, hedgings.

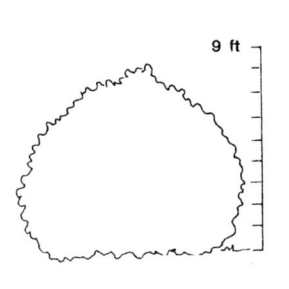

Registration: H. Baillon in *Monographie des Buxacées et des Stylocérées* 61, 1859. Invalid name: 'Macrophylla Rotundifolia.'
History: Most likely originated as an open-pollinated seedling somewhere in Europe. Believed to also have a yellow variegated form.
Bibliography:
Everett, T.H. *The New York Botanical Garden Ill. Encyc. of Hort.*, Vol. 2, 1981.
Flint, H.L. *Horticulture*, March 1987.
Hotte, A.C. *The Book of Shrubs*, 1931.
Krussman, G. *Manual of Cultivated Broad-leaved Trees and Shrubs*, Vol. I, A-D, 1984.
The Boxwood Bulletin, Vol. 16(1):15, July 1976/20(4):80, April 1981/21(4):66, April 1982/22(4):71, April 1983.
Known Locations: Arnold Arboretum, Buzzards Bay Garden Club, Royal Botanic Gardens, Edinburgh; Hillier's Arboretum, Secrest Arboretum, Washington Park Arboretum, U.S. National Arboretum, State Arboretum of Virginia.
Additional Information:
Culture and Care: Demonstrates no special cultural requirements.
Pests and Diseases: Indicates resistance to leaf miner, psyllid and mites in the more humid climates; no serious diseases.
Available in the commercial nursery trade.

Buxus sempervirens 'Salicifolia'

Size (25 yrs.):	Large – 6 to 7 feet high and 7 to 8 feet wide. A 30-year-old specimen measures 7 feet high and 9 feet wide.
Natural Form:	Mounded, billowy with spreading branches that tend to droop.
Annual Growth:	Fast – 3 to $3^1/_2$ inches in height and 4 to $4^1/_2$ inches in width.
Leaf Color:	Dark yellow-green.
Leaf Shape:	Elliptic and slightly revolute; acute tip; cuneate base.
Leaf Size:	Medium – $^1/_2$ to $^7/_{16}$ inch long and $^1/_4$ to $^1/_2$ inch wide.
Leaf Surface:	Glabrous and smooth.
Internodal Length:	Long – $^3/_8$ to $^7/_{16}$ inch.
Flowering Habit:	Floriferous and heavy fruiting.
Hardiness:	Zones 6 to 8.
Plant Use:	Specimen, grouping for background and area separations.

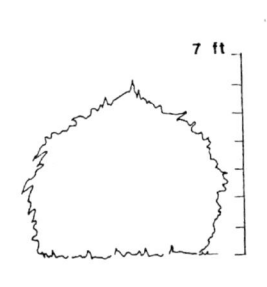

Registration: Hort. ex K. Koch in *Dendrologie*, v.2, pt. 2:476, 1872. Invalid names: 'Arborescens Salicifolia,' 'Ledifolia.'
History: Most likely originated as an open-pollinated seedling somewhere in Europe. Was listed in both French and German nursery catalogs in the late 1890s.
Bibliography:
Bailey, L.H. *Hortus Third*, 1976.
Missouri Botanical Garden Bulletin, Vol. LXXVI(3):8-10, May-June 1988.
The Boxwood Society of the Midwest Bulletin, April 1986.
The Boxwood Bulletin, Vol. 17(2):26, Oct. 1977/20(3):43,45, Jan. 1981/21(4):66, April 1982/27(3):65, Jan. 1988.
Known Locations: Arnold Arboretum, College of William and Mary, Kew Gardens, Missouri Botanical Garden, U.S. National Arboretum, State Arboretum of Virginia.
Additional Information:
Culture and Care: Demonstrates no special cultural requirements.
Pests and Diseases: Indicates resistance to leaf miner, psyllid and mites in the more humid climates; no serious diseases.
Available in the commercial nursery trade of North America and Europe.

Buxus sempervirens 'Salicifolia Elata'

Size (25 yrs.):	Large – 6 to 7 feet high and 11 to 12 feet wide. A 29-year-old specimen measures 7 feet high and 14 feet wide.
Natural Form:	Mounded, loose, open and twiggy habit.
Annual Growth:	Medium – 3 to 3½ inches in height and 5½ to 6 inches in width.
Leaf Color:	Medium green.
Leaf Shape:	Elliptic to lanceolate and slightly revolute; acute tip; cuneate base.
Leaf Size:	Large – ⁷/₈ to 1¹/₁₆ inches long and ⁵/₁₆ to ⁷/₁₆ inch wide.
Leaf Surface:	Glossy and smooth.
Internodal Length:	Long – ³/₈ to ½ inch.
Flowering Habit:	Moderate flowering and moderate fruiting.
Hardiness:	Zones 6 to 8.
Plant Use:	Specimen.

Registration: Catalog, F. Delauney, Angers, France, 1896.
History: Most likely originated as an open-pollinated seedling somewhere in Europe.
Bibliography:
Dallimore, W. *Holly, Yew and Box*, 1908.
The Boxwood Bulletin, Vol. 2(3):37, Jan. 1963.
Known Locations: Arnold Arboretum, College of William and Mary, Royal Botanic Gardens, Kew; Washington Park Arboretum, U.S. National Arboretum, State Arboretum of Virginia.
Additional Information:
Culture and Care: Demonstrates no special cultural requirements.
Pests and Diseases: Indicates resistance to leaf miner, psyllid and mites in the more humid climates; no serious diseases.
Not available in the commercial nursery trade.

Buxus sempervirens 'Ste. Genevieve'

Size (25 yrs.):	Large – 6 to 7 feet high and 8 to 9 feet wide. A 60-year-old specimen measures 10 feet high and 10 feet wide.
Natural Form:	Pyramidal – billowy.
Annual Growth:	Fast – 3 to 4 inches in height and 3½ to 4 inches in width.
Leaf Color:	Medium green.
Leaf Shape:	Uniformly lanceolate; acute tip; cuneate base.
Leaf Size:	Medium – $^{11}/_{16}$ to $^{13}/_{16}$ inch long and $^{5}/_{16}$ to $^{3}/_{8}$ inch wide.
Leaf Surface:	Glabrous – smooth.
Internodal Length:	Short – $^{3}/_{16}$ to $^{1}/_{4}$ inch.
Flowering Habit:	Moderate flowering and fruiting.
Hardiness:	Zones 5 to 8.
Plant Use:	Specimen, foundation planting, grouping for background and area separations, hedgings, topiary.

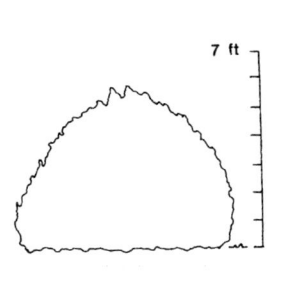

Registration: Mary A. Gamble in *The Boxwood Bulletin*, Vol. 11(1):15-16, July 1971.
History: The parent plant was discovered in 1934 in the Calvary Church Cemetery at Ste. Genevieve, Mo. Cuttings were planted at Gray Summit, Mo., for further testing. The origin of the parent plant is unknown but most likely was an open-pollinated seedling that originated in the eastern half of North America.
Bibliography:
Missouri Botanical Garden Bulletin, Vol. LXXVI(3):8-10, May-June 1988.
The Boxwood Society of the Midwest Bulletin, April 1986.
The Boxwood Bulletin, Vol. 11(1):1, July 1971/11(1):15-16, July 1971/16(4):54, April 1977/19(3):41, Jan. 1980/20(3):42,44, Jan. 1981/23(1):18, July 1983.
Known Locations: Buzzards Bay Garden Club, Longwood Gardens, Missouri Botanical Garden, U.S. National Arboretum, State Arboretum of Virginia.
Additional Information:
Culture and Care: Will tolerate siting in direct sun. Breaks dormancy late.
Pests and Diseases: Resistance to leaf miner, psyllid and mites.
Available in the commercial nursery trade.

Buxus sempervirens 'Suffruticosa'

Size (25 yrs.):	Dwarf – 2 to 2½ feet high and 2 to 2½ feet wide.
Natural Form:	Spherical to columnar – compact habit.
Annual Growth:	Slow – ¾ to 1¼ inches in height and ¾ to 1¼ inches in width.
Leaf Color:	Medium green.
Leaf Shape:	Obovate; retuse tip; cuneate base.
Leaf Size:	Medium – ⅝ to ¾ inch long and ⁷/₁₆ to ⅝ inch wide.
Leaf Surface:	Glossy – smooth.
Internodal Length:	Medium – ⁵/₁₆ to ⅜ inch.
Flowering Habit:	Rare flowering and fruiting.
Hardiness:	Zones 6 to 8.
Plant Use:	Specimen, grouping for background and area separations, edgings, hedgings.

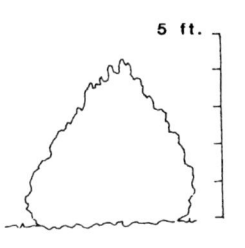

Registration: Linnaeus in *Species Plantarum*, 983,1753. Invalid names: 'Fruticosa,' 'Nana,' 'Rosmarinifolia Fruticosa,' 'Rosmarinifolia Minor,' 'Suffruticosa Nana,' 'Humilis.'

History: Most likely originated as an open-pollinated seedling somewhere in Europe. 'Suffruticosa' is probably one of the most often-planted cultivars. Erroneously called English, Dutch or French Box.

Bibliography:
Bailey, L.H. *Hortus Third*, 1976.
Bender, S. *Southern Living*, Nov. 1987.
Dirr, M.A. *Manual of Woody Landscape Plants*, 4th Ed., 1990.
Flint, H.L. *Horticulture*, March 1987.
Krussman, G. *Manual of Cultivated Broad-leaved Trees and Shrubs*, Vol. I, A-D, 1984.
Wright, M. *The Complete Handbook of Garden Plants*, 1984.
The Boxwood Bulletin, Vol. 1(3):26-27, April 1962/2(3):37, Jan. 1963/3(2):24, Oct. 1963/4(4):61-63, April 1965/5(3):44-47, Jan. 1966/6(3):37, Jan. 1967/7(2):17, Oct. 1967/9(1):8, July 1969/16(1):15, July 1976/18(2):cover, Oct. 1978/18(2):33-37, Oct. 1978/20(4):80, April 1981/20(4):64-66, April 1981/21(3):45, Jan. 1982/ 21(4):66, April 1982/22(4):70, April 1983/27(3):65, Jan. 1988.

Known Locations: Arnold Arboretum, Brooklyn Botanic Garden, College of William and Mary, Hillier's Arboretum, Longwood Gardens, Oxford Botanic Gardens, Tennessee Botanic Gardens, The Hermitage, U.S. National Arboretum, State Arboretum of Virginia.

Additional Information:

Culture and Care: Quite fussy about its cultural care and requires annual maintenance. Keep the dead leaves cleaned out of the center of the plant to prevent adventitious rooting and fungus growth in the more humid climates. Since it is such a compact-growing plant, judiciously prune or pluck to keep it opened up, particularly in the top section, to provide ventilation and sunlight. The increase in stem leaves will help support the feeder roots. It does not take well to shearing; prune for shape.

Pests and Diseases: Resistance to leaf miner. Attraction to psyllid infestation.

Available in the commercial nursery trade.

Buxus sempervirens 'Tennessee'

Size (25 yrs.):	Medium – 4 to 4½ feet high and 5 to 5½ feet wide.
Natural Form:	Pyramidal.
Annual Growth:	Medium – 2 to 2½ inches in height and 2½ to 2¾ inches in width.
Leaf Color:	Medium green.
Leaf Shape:	Elliptic to ovate; acute tip; cuneate base.
Leaf Size:	Medium – ⅝ to ⅞ inch long and ⁵⁄₁₆ to ⅜ inch wide.
Leaf Surface:	Glossy and smooth.
Internodal Length:	Medium – ⁵⁄₁₆ to ⁷⁄₁₆ inch.
Flowering Habit:	Not observed.
Hardiness:	Zones 5 to 8.
Plant Use:	Specimen, foundation planting, grouping for background and area separations, hedgings.

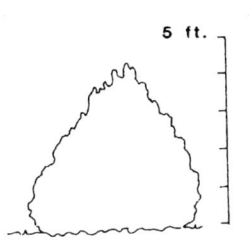

Registration: Not registered.
History: Martin Bagby of Pacific, Mo., provided the cuttings to The Boxwood Society of the Midwest. The original cuttings were from a plant in Tennessee and brought to Gray Summit, Mo., in 1944 by Walter Taylor.
Bibliography:
The Boxwood Society of the Midwest Bulletin, Vol. 8(1):4, July 1983.
Known Locations: Not documented
Additional Information:
Culture and Care: Demonstrates no special cultural requirements.
Pests and Diseases: Indicates resistance to leaf miner, psyllid and mites in the more
 humid climates; no serious diseases.
Not available in the commercial nursery trade.

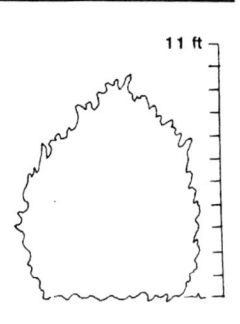

Buxus sempervirens 'Undulifolia'

Size (25 yrs.):	Large – 10 to 11 feet high and 8 to 9 feet wide. A 29-year-old specimen plant measures 12 feet high and 9^1/$_2$ feet wide.
Natural Form:	Arboreal, upright stiff growth, loose and open habit.
Annual Growth:	Fast – 4^1/$_2$ to 5 inches in height and 3^1/$_2$ to 4 inches in width.
Leaf Color:	Dark green.
Leaf Shape:	Lanceolate tending toward ovate and slightly revolute; obtuse tip with some being retuse; cuneate base.
Leaf Size:	Large – ^{3}/$_4$ to 1 inch long and ^{7}/$_{16}$ to ^{5}/$_8$ inch wide.
Leaf Surface:	Glabrous and smooth.
Internodal Length:	Medium – ^{1}/$_4$ to ^{3}/$_8$ inch.
Flowering Habit:	Floriferous and heavy fruiting.
Hardiness:	Zones 6 to 8.
Plant Use:	Specimen.

Registration: *Kew Handlist of Trees and Shrubs*, 270, 1902.
History: Most likely originated as an open-pollinated seedling somewhere in Europe.
Bibliography:
Dallimore, W. *Holly, Yew and Box*, 1908.
Krussman, G. *Manual of Cultivated Broad-leaved Trees and Shrubs*, Vol. I, A-D, 1984.
Wyman, D. *Wyman's Gardening Encyclopedia*, New Expanded 2nd Ed., 1986.
The Boxwood Bulletin, Vol. 2(3):37, Jan. 1963.
Known Locations: Arnold Arboretum, Royal Botanic Gardens, Kew; Washington Park Arboretum, U.S. National Arboretum, State Arboretum of Virginia.
Additional Information:
Culture and Care: Demonstrates no special cultural requirements.
Pests and Diseases: Indicates resistance to leaf miner, psyllid and mites in the more humid climates; no serious diseases.
Not available in the commercial nursery trade.

Buxus sempervirens 'Vardar Valley'

Size (25 yrs.):	Medium – 2½ to 3 feet high and 7 to 8 feet wide. A 30-year-old specimen measures 3¼ feet high and 10 feet wide.
Natural Form:	Mounded.
Annual Growth:	Slow – 1 to 1½ inches in height and 3 to 4 inches in width.
Leaf Color:	Medium green with bluish hue.
Leaf Shape:	Rotund to obovate; obtuse tip, some retuse; cuneate base.
Leaf Size:	Large – ¾ to 1⅛ inches long and ⁵/₁₆ to ⅝ inch wide.
Leaf Surface:	Glabrous – smooth.
Internodal Length:	Long – ⅜ to ⁷/₁₆ inch.
Flowering Habit:	Floriferous and heavy fruiting.
Hardiness:	Zones 5 to 8.
Plant Use:	Specimen, foundation planting, grouping for background and area separations, hedgings.

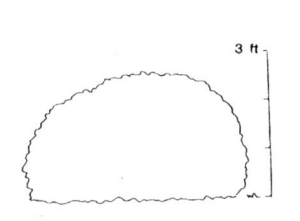

Registration: D. Wyman in *Arnoldia*, 17(7):42-44, 1957.
History: Collected by Dr. Edgar Anderson in 1934 from the Balkans. Accessioned at Arnold Arboretum and labeled "352-35 B. semp. 'VV' E.A. 133 Treska Gorge, Skoplje, 40 cutts." Commercially released by Kingsville Nurseries, Kingsville, Md.
Bibliography:
Bender, S. *Southern Living*, Nov. 1987.
Dirr, M.A. *Manual of Woody Landscape Plants*, 4th Ed., 1990.
Flint, H.L. *Horticulture*, March 1987.
Gamble, M.A. *Flower and Garden*, March 1988.
Wyman, D. *Wyman's Gardening Encyclopedia*, Expanded 2nd Ed., 1986.
Missouri Botanical Garden Bulletin, Vol. LXXVI(3):8-10, May-June 1988.
The Boxwood Society of the Midwest Bulletin, April 1986.
The Boxwood Bulletin, Vol. 9(1):8, July 1969/14(4):59,62, April 1975/16(1):15, July 1976/17(1):9-10, July 1977/18(2):43, Oct. 1978/20(3):41,44, Jan. 1981/20(4):80, April 1981/21(4):66, April 1982/23(1):19, July 1983/24(2):cover,43-44, Oct. 1984/ 24(2):52-53, Oct. 1984/27(2):37, Oct. 1987/27(3):65, Jan. 1988.

Known Locations: Arnold Arboretum, Buzzards Bay Garden Club, College of William and Mary, Hillier's Arboretum, Longwood Gardens, Missouri Botanical Garden, Secrest Arboretum, U.S. National Arboretum, State Arboretum of Virginia.

Additional Information:
Cultural Care: Tolerates siting in direct sun quite well, but protect from winter winds. Breaks dormancy late.
Pests and Diseases: Resistance to leaf miner and mites. Attraction to psyllid infestation.
Available in the commercial nursery trade.

Buxus sempervirens 'Welleri'

Size (25 yrs.):	Medium – 3 to 4 feet high and 4 to 4½ feet wide.
Natural Form:	Pyramidal.
Annual Growth:	Medium – 1½ to 2 inches in height and 1½ to 2 inches in width.
Leaf Color:	Medium green.
Leaf Shape:	Elliptic – slightly revolute; acute tip; cuneate base.
Leaf Size:	Medium – ⅝ to 1 inch long and ⁵/₁₆ to ⅜ inch wide.
Leaf Surface:	Glabrous – smooth.
Internodal Length:	Medium – ⁵/₁₆ to ⅜ inch.
Flowering Habit:	Sparse flowering and fruiting.
Hardiness:	Zones 5 to 8.
Plant Use:	Specimen, foundation planting, grouping for background and area separations, hedgings.

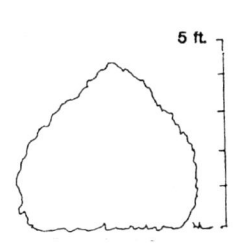

Registration: Catalog, Weller Nursery Company, Holland, Mich., 1945.
History: Most likely originated as an open-pollinated seedling somewhere in Michigan.
Bibliography:
Bailey, L.H. *Hortus Third*, 1976.
Bender, S. *Southern Living*, Nov. 1987.
Dirr, M.A. *Manual of Woody Landscape Plants*, 4th Ed., 1990.
Flint, H.L. *Horticulture*, March 1987.
Gamble, M.A. *Flower and Garden*, March 1988.
Wyman, D. *Wyman's Gardening Encyclopedia*, New Expanded 2nd Ed., 1986.
The Boxwood Society of the Midwest Bulletin, April 1986.
The Boxwood Bulletin, Vol. 17(1):9, July 1977/21(4):66, April 1982/27(2):37, Oct. 1987.
Known Locations: Arnold Arboretum, Buzzards Bay Garden Club, U.S. National Arboretum, State Arboretum of Virginia.
Additional Information:
Culture and Care: Another of my favorites for small properties. Tolerates siting in direct sun quite well, but protect from winter winds. Breaks dormancy late.
Pests and Diseases: Resistance to leaf miner, psyllid and mites.
Available in the commercial nursery trade.

Buxus sempervirens 'West Ridgeway'

Size (25 yrs.):	Small – 2½ to 3 feet high and 2½ to 3 feet wide.
Natural Form:	Spherical.
Annual Growth:	Slow – 1 to 1½ inches in height and 1 to 1½ inches in width.
Leaf Color:	Medium green.
Leaf Shape:	Obovate; acute tip tending toward rotund, slightly retuse; cuneate base.
Leaf Size:	Small – ½ to ⅝ inch long and ¼ to 5/16 inch wide.
Leaf Surface:	Glabrous and smooth.
Internodal Length:	Medium – 5/16 to ⅜ inch.
Flowering Habit:	Not observed.
Hardiness:	Zones 6 to 8.
Plant Use:	Specimen, grouping for background and area separations, edgings, hedgings.

3 ft

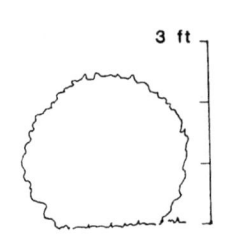

Registration: Not registered.
History: The parent clone was found in the West Ridgeway Cemetery in the town of Ridgeway, Orleans County, western New York. Dr. Benjamin Blackburn of Willowwood Arboretum in 1986 said that he had watched and admired it for many years and believed the plant may have dated to the early 1800s and possibly 1790s. Both Dr. Blackburn and Dr. Henry Skinner of the National Arboretum believed the plant might well have been from a mutation of *Buxus sempervirens* 'Suffruticosa.'
Bibliography:
The Boxwood Bulletin, Vol. 26(1):18, July 1986.
Known Locations: U.S. National Arboretum, State Arboretum of Virginia.
Additional Information:
Culture and Care: Demonstrates no special cultural requirements.
Pests and Diseases: Indicates resistance to leaf miner, psyllid and mites in the more humid climates; no serious diseases.
Available in the commercial nursery trade.

Buxus sempervirens 'Woodland'

Size (25 yrs.):	Medium – 5½ to 6 feet high and 6 to 6½ feet wide.
Natural Form:	Pyramidal and billowy.
Annual Growth:	Medium – 3 to 3¼ inches in height and 2½ to 3 inches in width.
Leaf Color:	Dark green.
Leaf Shape:	Lanceolate to obovate; acute to obtuse tip; cuneate base.
Leaf Size:	Large – ¾ to 1 inch long and ½ to ⅝ inch wide.
Leaf Surface:	Glabrous and smooth.
Internodal Length:	Medium – ½ to ⅝ inch.
Flowering Habit:	Not observed.
Hardiness:	Zones 6 to 8.
Plant Use:	Specimen, foundation planting, grouping for background and area separations, hedgings.

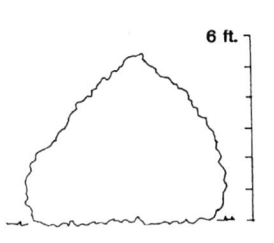

Registration: Not registered.
History: Believed to have originated as an open-pollinated seedling in Ontario, Canada.
Bibliography: Not documented
Known Locations: U.S. National Arboretum, State Arboretum of Virginia.
Additional Information:
Culture and Care: Demonstrates no special cultural requirements.
Pests and Diseases: Indicates resistance to leaf miner, psyllid and mites in the more humid climates; no serious diseases.
Not available in the commercial nursery trade.

Buxus sempervirens 'Yorktown'

Size (25 yrs.):	Medium – 5 to 5½ feet high and 5 to 5½ feet wide.
Natural Form:	Generally pyramidal, loose and open habit.
Annual Growth:	Medium – 2½ to 3 inches in height and 2½ to 3 inches in width.
Leaf Color:	Medium yellow-green.
Leaf Shape:	Lanceolate tending toward obovate; acute tip; cuneate base.
Leaf Size:	Large – ⅞ to 1 inch long and ⅜ inch wide.
Leaf Surface:	Glossy and smooth.
Internodal Length:	Medium – ½ to ⅝ inch.
Flowering Habit:	Not observed.
Hardiness:	Zones 6 to 8.
Plant Use:	Specimen.

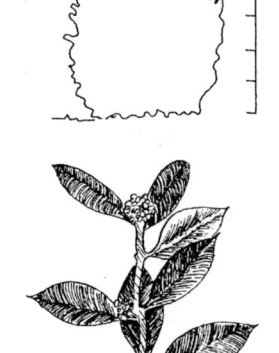

Registration: Not registered.
History: The origin of the plant is unknown. However, Dr. J.T. Baldwin, Jr. of the College of William and Mary selected and named the plant.
Bibliography: Not documented
Known Locations: State Arboretum of Virginia.
Additional Information:
Culture and Care: Demonstrates no special cultural requirements.
Pests and Diseases: Indicates resistance to leaf miner, psyllid and mites in the more humid climates; no serious diseases.
Not available in the commercial nursery trade.

Buxus sempervirens 'Zehtung'

Size (25 yrs.):	Expected to be a medium-sized shrub about 4 to 4½ feet high and 3 to 3½ feet wide.
Natural Form:	Generally pyramidal and somewhat billowy based upon a 15-year-old plant.
Annual Growth:	Medium – approximately 2 to 2½ inches in height and 1½ to 2 inches in width.
Leaf Color:	Medium green.
Leaf Shape:	Uniformly lanceolate and slightly revolute; acute tip; cuneate base.
Leaf Size:	Medium – ⅝ to ¾ inch long and ¼ to 5/16 inch wide.
Leaf Surface:	Glabrous, smooth and leathery.
Internodal Length:	Medium – ⅜ to 7/16 inch.
Flowering Habit:	Not observed.
Hardiness:	Zones 5 to 8.
Plant Use:	Specimen.

Registration: Not registered.
History: Believed to have originated as an open-pollinated seedling of *Buxus sempervirens*, named and distributed by an unknown nursery in Ohio.
Bibliography:
Dirr, M.A. *Manual of Woody Landscape Plants,* 4th Ed., 1990.
Known Locations: Missouri Botanical Garden, State Arboretum of Virginia.
Additional Information:
Culture and Care: Demonstrates no special cultural requirements.
Pests and Diseases: Indicates resistance to leaf miner, psyllid and mites in the more humid climates; no serious diseases.
Available in the commercial nursery trade.

Selected Species, Varieties, Cultivars

Buxus sempervirens – Partial Descriptions

Cultivars with insufficient data for a reasonable plant description:

'Anderson 350-35' – Inventory U.S. National Arboretum (4213).

'Angustifolia Variegata Punctulata' – H. Baillon in *Monographie des Buxacées et des Stylocérées*, 61, 1859.

'Arborescens Argentea' – J. Loudon in *Arboretum et Fruticum Britannicum*, III:1333, 1893.

'Arborescens Aurea Acuminata' – H. Baillon in *Monographie des Buxacées et des Stylocérées*, 60, 1859.

'Arborescens Aurea Punctulata' – H. Baillon in *Monographie des Buxacées et des Stylocérées*, 60, 1859.

'Arborescens Decussata' – *Kew Handlist of Trees and Shrubs*, 269, 1925.

'Argentea Nova' – Catalog, V. Gauntlett, Chittingfold, Surrey, England, 1930.

'Balkan' – Plant List, H.E. Nursery, 1302 Union St., Litchfield, Ill., 1983.

'Bass' – M.A. Dirr, *Manual of Woody Landscape Plants*, 4th Ed. "Medium-sized, densely branched form with slightly larger foliage than 'Inglis' and the dark green color is retained through winter."

'BEF 287' – Inventory, State Arboretum of Virginia.

'Bowles Blue' – Inventory, Royal Horticultural Society, 1988.

'Broman' – Sheridan Nurseries, Toronto, Canada. Selected in 1936. Inventory, Arnold Arboretum.

'California' – Catalog, Moreau Landscape Nursery, East Colts Neck, N.J., 1985.

'Caucasica' – Hort. ex K. Koch in *Dendrologie*, v. 2, pt. 2:476, 1862.

'Christiansen' – Catalog, Cary Brothers Nursey, Shrewsbury, Mass., 1957.

'Columnaris' – Catalog, Visser's Nurseries, 132-9 Merrick Blvd., Springfield Gardens, Long Island, N.Y., 1960. Columnar; *Hortus III*, "erect, columnar." Inventory, Longwood Gardens.

'Compacta' – Catalog, Charles Dietriche, Angers, France, 1953.

'Conica' – Catalog, Siebenthaler Nurseries, Dayton, Ohio 136:10, 1938. *Hortus Third*, "erect, conical."

'Crispa' – Hort. ex C. Koch in *Dendrologie*, v. pt. 2:476, 1872.

'Croni' – Catalog, Monroe Nurseries, Monroe, Mich., 1955. *Hortus Third*: "strong growing, hardy." *The Boxwood Bulletin*, Vol. 21(4):64-65, April 1982. "Grown for four years. Planted in three locations. Two plants killed during two successive severe winters with temperatures to -17° F. These two plants had also been planted so deep for mortality could have been a combination of poor planting and low temperatures. Other plantings have only experienced -14° F. Plants were 22 inches high with 15- to 20-inch spread. Half the tops killed back at -14° F. Plants recovered during following growing season and are presently (Nov. 1981) upright growing plants with a good green color. Plants haven't been growing long enough to definitely evaluate, but it appears that they will at best be only marginally hardy on protected sites (Zone 5)."

'Cucullata' – Hort. ex C. Koch in *Dendrologie*, v. 2, pt. 2:476, 1872. *Handbuch der Laubholzkunde*, 3(82), 1893. "Mit in der Mitte vertieften."

'Elegans' – L.H. Bailey in *Standard Cyclopedia of Horticulture*, 601, 1914. *Hortus Third*, "leaves white-variegated, oblong." G. Krussman in *Manual of Cultivated Broad-leaved Trees and Shrubs*: "weak grower; upright; leaves narrow, often somewhat deformed, white margined (= Elegantissima). Frequently cultivated in French nurseries."

'Fairview' – Catalog, Eastern Shore Nurseries, Inc., Easton, Md., p. 47, 1947. "A large leaved variety; leaves ovate but about twice the size of the type; named after the old Colonial estate near Easton." *The Boxwood Bulletin*, Vol. 22(4):71, April 1983.

'Falkner' – Catalog, Firma C. Esveld, Boskoop, Holland, 1987.

'Fastigiata Hardwickensis' – Catalog, Kingsville Nurseries, Kingsville, Md. Believed to have come to Kingsville Nurseries from Eastern Shore Nurseries, Easton, Md., which had received it from England. Inventory, Washington Park Arboretum, 1983.

'Fiesta' – Catalog, Firma C. Esvold, Boskoop, Holland, 1987.

'Glauca Marginata Aurea' – Catalog, F. Delaunay, Angers, France, 1910.

'Globosa' – Catalog, Siebenthaler Nurseries, Dayton, Ohio 136:10, 1938.

'Golden Frimley' – Inventory, Royal Botanic Gardens, Kew, London, England.

'Grandifolia' – J. Mueller, Arg. in *De Candolle Prodromus* 16(1):19, 1869.

'Grand Rapids' – Catalog, Light's Tree Co., Richland, Mich., 12:14, 1948. "Hardy."

'Gray Summit' – Inventory, Secrest Arboretum, Wooster, Ohio, 1982. "Has been outplanted for one year. Survived first winter with temperatures to -14° F. with long periods of below zero in good condition. It is currently an upright growing plant 18 inches tall with a 10-inch spread and of a good green color to the foliage. Has not been grown long enough in Arboretum to determine hardiness but shows promise." *The Boxwood Bulletin*, Vol. 21(4):65, April 1982.

'Green Beauty' – Catalog, Eastern Shore Nurseries, Easton, Md., 1964. "A new variety with smaller leaves; darker green and more compact." M.A. Dirr, *Manual of Woody Landscape Plants,* 3rd Ed., 1983, "upright form with year-round thick, dense, dark green foliage." Also see *B. microphylla* variety *japonica* "Green Beauty."

'Hamilton' – D. Wyman, *Wyman's Gardening Encyclopedia*, New Exp. 2nd Ed., 1986, "hardy." This plant could very well be *B. sempervirens* 'Northern Find,' which originated in that part of Ontario and is often referred to as the Hamilton variety.

'Handsworthiensis Candelabra' – Catalog, Kingsville Nurseries, Kingsville, Md., 1967. Inventory, Washington Park Arboretum, 1983; U.S. National Arboretum, 1988.

'Handsworthii Aurea' – Catalog, Visser's Springfield Gardens, Merrick Road, Springfield, Long Island, N.Y., 1945.

'Hardy Michigan' – Catalog, John Vermeulen and Son, Inc., Neshanic Station, N.J., 1959. "Relatively hardy."

'Harmony Grove' – D. Wyman, *American Nurseryman* 107(7):57, 1963. "A new selection being grown by the Kelsy Highlands Nurseries, East Boxford, Mass." "Originated at Harmony Grove Cemetery, Salem, Mass."

'Hendersonii' – Catalog, Lindley Nurseries, Greensboro, N.C., 1958.
'Heterophylla' – V. Veillard in *Duhamel, Traité des Arbres et Arbrisseaux*, ed. augm. 1:82, 1835.
'Hillsboro' – Rocknoll Nursery 9210 U.S. 50, Hillsboro, Ohio, 45133, 187.
'Hirsholmi' – Kobenhavens Univ. Botanisk. 1986.
'Jensen' – Catalog, Moreau Landscape Nursery, Inc., Colts Neck, N.J. Listed in the inventory of the Brooklyn Botanic Garden.
'Kirkham' – Inventory, Missouri Botanical Garden.
K-2 – Inventory, U.S. National Arboretum (9409-C).
K-8 – Inventory, U.S. National Arboretum (9415-C)
K-12 – Inventory, U.S. National Arboretum (9419-C).
K-16 – Inventory, U.S. National Arboretum (9422).
K-19 – Inventory, U.S. National Arboretum (9425).
K-23 – Inventory, U.S. National Arboretum (9429-C).
K-28 – Inventory, U.S. National Arboretum (9434-C).
K-32 – Inventory, U.S. National Arboretum (9437-C).
K-34 – Inventory, U.S. National Arboretum (9439).
K-36 – Inventory, U.S. National Arboretum (9453-C).
K-37 – Inventory, U.S. National Arboretum (9441).
K-47 – Inventory, U.S. National Arboretum (9451-C).
K-52 – Inventory, U.S. National Arboretum (9457).
K-65 – Inventory, U.S. National Arboretum (9470).
K-70 – Inventory, U.S. National Arboretum (9475).
K-74 – Inventory, U.S. National Arboretum (9479).
K-89 – Inventory, U.S. National Arboretum (9494-C).
K-91 – Inventory, U.S. National Arboretum (9496).
K-94 – Inventory, U.S. National Arboretum (9498-C).
K-100 – Inventory, U.S. National Arboretum (9502-C).
K-107 – Inventory, U.S. National Arboretum (9509).
K-112 – Inventory, U.S. National Arboretum (9512-C).
K-114 – Inventory, U.S. National Arboretum (9514-C).
K-118 – Inventory, U.S. National Arboretum (9518-C).
K-119 – Inventory, U.S. National Arboretum (9519-C).
K-128 – Inventory, U.S. National Arboretum (9525-C).
K-130 – Inventory, U.S. National Arboretum (9527).
K-133 – Inventory, U.S. National Arboretum (9531-C).
K-134 – Inventory, U.S. National Arboretum (9532-C).
K-140 – Inventory, U.S. National Arboretum (9538).
K-144 – Inventory, U.S. National Arboretum (9541).
K-146 – Inventory, U.S. National Arboretum (9543-C).
'Latifolia Aurea' – Inventory, Royal Botanic Gardens, Edinburgh, Scotland, 1988.

'Latifolia Pendula' – Inventory, Hillier's Arboretum.

'Latifolia Rotundifolia' – Catalog, Kingsville Nurseries, Kingsville, Md. Inventory, U.S. National Arboretum (29699).

'Lynnhaven' – Catalog, Greenbrier Farms, Inc., Norfolk, Va., 1922. "Broad pyramidal growth."

'Maculata Pendula' – Inventory, Washington Park Arboretum, 1983; U.S. National Arboretum, 1988.

'Maplewood' – Catalog, Bobbink Nurseries, Freehold, N.J., 1987.

'Meyers Columnar' – Inventory, State Arboretum of Virginia (17821-85).

'Minima' – Beissner, Schelle and Zabel in *Handbuch der Laubholz-Benennung*, 283, 1903.

'Minima Glauca' – Catalog, Charles Dietriche, Angers, France, 1892.

'Minor-aureo' – R. Weston in *Botanicus Universalis*, 1:31, 1770.

'Mucronata' – Hortul. ex Baillon, H. in *Monographie des Buxacées et des Stylocérées*, 62, 1859.

Mulstead selection – *The Boxwood Bulletin*, Vol. 21(4):65, April 1982. "Grown for 14 years. Plants grew exceptionally well with average yearly growth of more than 6 inches a year at temperatures to -12º F. During three successive winters when temperatures dropped to -20º F. and -17º F., considerable top killing on plants growing on site somewhat exposed to sun and wind. Plants were killed back from 65 inches tall to 30 inches tall. These boxwoods are currently in very poor condition. Bark killed on south side of main stems. Mulstead selections on well-protected site with light high shade suffered up to 20 percent kill at -20º F. but have completely recovered and are presently 76 to 80 inches tall. They have a good green color and are in good condition. Mulstead selection of Common Box have been marginally hardy in the arboretum (Secrest) depending upon site. Have done well on very protected site. They have grown poorly on average protected sites with some exposure."

'Nana' – Inventory, Royal Botanic Gardens, Kew.

'Nigricans' – P. Corbelli in *Dizionario di Floricultura* 232, 1873.

'Northern Beauty' – M.A. Dirr in *Manual of Woody Landscape Plants*, 3rd Ed., 1983. "A form that survived the terrible winters of 1976-77, 77-78 with no foliage burn."

'Northern New York' – Inventory, Beal-Garfield Botanic Garden, East Lansing, Mich., 1960; Secrest Arboretum, Wooster, Ohio, 1982; Washington Park Arboretum, 1983; U.S. National Arboretum (29701). *The Boxwood Bulletin*, Vol. 21(4):65, April 1982. "Grown for 14 years. Grew well until three successive winters with temperatures of -20º F. and -17º F. Plants on site somewhat exposed to winds and winter sun were completely killed at -20º F. Boxwood on more protected site were killed to the ground although they were completely covered with snow. Plant has resprouted and is presently 6 inches tall and in poor condition. Northern New York Common Box has not been hardy in the (Secrest) Arboretum at temperatures below -12º F."

'Ohio' – Catalog, Bobbink Nurseries, Freehold, N.J., 1987.

'Oleaefolia' – L.H. Bailey in *Standard Cyclopedia of Horticulture*, 601, 1914. Upright habit, leaves oblong resembling those of an olive.

Selected Species, Varieties, Cultivars

'Ponteyi Variegata' – Inventory, Kew Gardens, 1988.

'Prizren' – Inventory, Morris Arboretum, 1986. *The Boxwood Bulletin,* Vol. 26(2):39, Oct. 1986. "Among the Anderson cultivars at Morris Arboretum (these are descended from seed that Dr. Edgar Anderson had sent back from the Balkans in the 1930s), one named 'Prizren' stands out. It grows 6 to 8 inches a year, and at 38 years is perhaps 20 feet tall. Its width is pleasingly proportioned to its height, and it is obviously hardy and vigorous."

'Pyramidalis Variegatis' – Catalog, Baudriller Nurseries, Angers, France, 1880.

'Rotundifolia Aurea' – L. Dippel in *Handbuch der Laubholzkunde,* 3:82, 1893.

'Rotundifolia Aureo-variegata' – Beissner, Schelle and Zabel in *Handbuch der Laubholz-Benennung,* 284, 1903.

'Rotundifolia Glauca' – Catalog, Charles Dietriche, Angers, France, 1892.

'Rotundifolia Maculata' – F. Meyer in *Plant Explorations,* ARS 34-9:113b, 1959.

'Rotundifolia Minor' – Beissner, Schelle and Zabel in *Handbuch der Laubholz-Benennung,* 284, 1903.

'Schmidt' – *The Boxwood Bulletin,* Vol. 16(1):15, July 1976/21(4):66, April 1982. "Grown for 12 years. Grew exceptionally well until winter with -20º F. when 60 percent of twigs were winter-killed. After two following severe winters additional damage occurred. Southwest exposed side of one plant killed and top of second plant. Plants have recovered and are currently 36 and 50 inches tall with dense green foliage. They have a spread of 30 inches. Schmidt boxwood has been marginally hardy in the arboretum (Secrest) on protected sites."

'Semi-elata' – Catalog, Charles Dietriche, Angers, France, 1892.

'Semperaurea' – B. Wagenknecht in *The Boxwood Bulletin,* Vol. 7(1):1, 1967.

'Serbian Blue' – *The Boxwood Bulletin,* Vol. 14(4):61, April 1975. It is possible that *B. sempervirens* 'Nish' (see p. 158) and 'Serbian Blue' are the same seed lot received from the Balkans by Dr. Edgar Anderson in 1935.

'Strassner' – Inventory, U.S. National Arboretum accession # (34198).

'Subglobosa' – Beissner, Schelle and Zabel in *Handbuch der Laubholz-Benennung,* 283, 1903.

'Suffruticosa Alba Marginata' – Catalog, Brimfield Nurseries, Wethersfield, Conn., 1955.

'Suffruticosa Aurea' – H. Baillon in *Monographie des Buxacées et des Stylocérées,* 61, 1859.

'Suffruticosa Aureo-marginata' – Beissner, Schelle and Zabel in *Handbuch der Laubholz-Benennung,* 284, 1903.

'Suffruticosa Crispa' – Beissner, Schelle and Zabel in *Handbuch der Laubholz-Benennung,* 284, 1903.

'Suffruticosa Glauca' – Beissner, Schelle and Zabel in *Handbuch der Laubholz-Benennung,* 284, 1903.

'Suffruticosa Maculata' – Beissner, Schelle and Zabel in *Handbuch der Laubholz-Benennung,* 284, 1903.

'Suffruticosa Variegata' – *Journal of the Royal Horticultural Society,* 18:22, 1895.

'Tenuifolia' – Hortul. ex H. Baillon in *Monographie des Buxacées et des Stylocérées,* 61, 1859.

'Thymifolia' – Beissner, Schelle and Zabel in *Handbuch der Laubholz-Benennung*, 284, 1903. M. Wright, *The Complete Handbook of Garden Plants*, 1984.

'Thymifolia Variegata' – *Journal of The Royal Horticultural Society*, 18:82, 1895.

'Variegata' – Hort. ex Steud. Nom. ed. II.i.242. *Hortus Third*, "leaves white or yellowish variegated." Catalog, Firma C. Esveld, Boskoop, Holland, 1987.

'Washington-Missouri' – Inventory, U.S. National Arboretum accession # (33795).

'William Borek' – Inventory, Arnold Arboretum, 1986.

'Wooster No. 1' – *The Boxwood Bulletin*, Vol. 21(4):66, April 1982. "Has been set out in six separate locations since original selection was made in 1948. Plants grew well until series of extra severe winters with temperatures from -17° F. to -20° F. began with winter of 1976-77. First winter with temperatures of -20° F. resulted in 60 percent top kill. After three successive severe winters, plants were killed to snow line. They were one third the height they had been before the advent of extremely severe winters. Plants 30 inches high set in the fall of 1980 were killed to the snow line during the winter of 1980-81 with temperatures to -14° F. Plants are currently in fair condition having recovered in a great degree from winter kill. Foliage is currently bronze from early October 1981 freeze. Wooster No. 1 Boxwood is at best only marginally hardy in the Arboretum (Secrest) on well protected sites."

Buxus sinica variety *insularis* 'Justin Brouwers'

Size (25 yrs.):	Small – 2 to 2½ feet high and 3 to 3½ feet wide. A 30-year-old specimen measures 2¾ feet high and 4 feet wide.
Natural Form:	Mounded.
Annual Growth:	Slow – 1 to 1½ inches in height and 1½ to 1¾ inches in width.
Leaf Color:	Dark green – holds color well during winter months.
Leaf Shape:	Lanceolate; acute tip; cuneate base.
Leaf Size:	Small – ⅝ to ¾ inch long and ¼ to 5/16 inch wide.
Leaf Surface:	Glabrous – smooth and shiny.
Internodal Length:	Short – 3/16 to ¼ inch.
Flowering Habit:	Not observed.
Hardiness:	Zones 6 to 8.
Plant Use:	Specimen, grouping for background and area separations, edgings, hedgings.

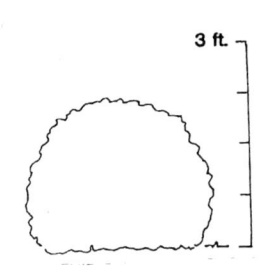

Registration: P.D. Larson, *The Boxwood Bulletin*, Vol. 29(1):3, July 1989.
History: Discovered as a seedling of *Buxus sinica* variety *insularis* grown and selected by Justin B. Brouwers, former landscape superintendent at Colonial Williamsburg, Va. It has been said that he liked this seedling so much it was placed around the grave of a favorite cat and some local residents still refer to it as Cat's Grave Seedling.
Bibliography:
The Boxwood Bulletin, Vol. 23(1):14, July 1983/27(1):32, July 1987.
Known Locations: Buzzards Bay Garden Club, U.S. National Arboretum, State Arboretum of Virginia.
Additional Information:
Culture and Care: An outstanding plant. Tolerates siting in direct sun and holds its color well. Breaks dormancy late.
Pests and Diseases: Resistance to leaf miner, psyllid and mites.
Available in the commercial nursery trade.

Buxus sinica variety *insularis* 'Nana'

Size (25 yrs.):	Small – 2^1/$_2$ to 3 feet high and 3 to 3^1/$_2$ feet wide.
Natural Form:	Mounded – compact.
Annual Growth:	Slow – 1 to 1^1/$_4$ inches in height and 1^1/$_4$ to 1^1/$_2$ inches in width.
Leaf Color:	Medium yellow-green.
Leaf Shape:	Lanceolate – some ovate; obtuse tip, some retuse; cuneate base.
Leaf Size:	Small – ^{1}/$_2$ to ^{5}/$_8$ inch long and ^{3}/$_{16}$ to ^{5}/$_{16}$ inch wide.
Leaf Surface:	Glabrous – smooth.
Internodal Length:	Medium – ^{1}/$_4$ to ^{3}/$_8$ inch.
Flowering Habit:	Floriferous.
Hardiness:	Zones 6 to 8.
Plant Use:	Specimen, grouping, edgings, hedgings.

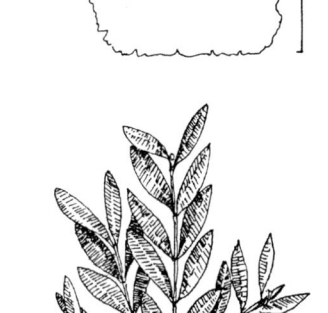

Registration: Beissner, Schelle, and Zabel in *Handbuch der Laubholz-Benennung,* 283, 1903. BB, Vol. 34, (1):32, July, 1994.
History: Not documented
Bibliography: Not documented
Known Locations: Buzzards Bay Garden Club, Washington Park Arboretum, U.S. National Arboretum, State Arboretum of Virginia.
Additional Information:
Culture and Care: Quite fussy about its cultural care as it matures. Prefers dappled
    shade. Keep opened up, particularly in the top section, to provide ventilation and
    sunlight to penetrate and increase stem leaves to help support the feeder roots.
Pests and Diseases: Resistance to leaf miner and mites. Attraction to psyllid infestation.
Not available in the commercial nursery trade.

Buxus sinica variety *insularis* 'Pincushion'

Size (25 yrs.):	Small – 3 to 3½ feet high and 4 to 4½ feet wide.
Natural Form:	Mounded – loose open habit.
Annual Growth:	Medium – 1½ to 1¾ inches in height and 2 to 2½ inches in width.
Leaf Color:	Medium green with yellow undertone.
Leaf Shape:	Rotund; obtuse tip, some retuse; cuneate base.
Leaf Size:	Small – ⅝ to ¾ inch long and 5/16 to 7/16 inch wide.
Leaf Surface:	Glabrous – smooth.
Internodal Length:	Medium – ¼ to ⅜ inch.
Flowering Habit:	Moderate flowering and fruiting.
Hardiness:	Zones 5(protected) to 8.
Plant Use:	Specimen, grouping for background and area separations, edgings, hedgings.

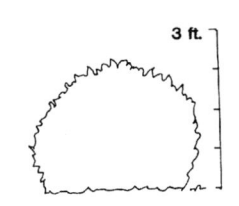

Registration: B. Wagenknecht in *The Boxwood Bulletin,* Vol. 7(1):1, July 1967.
History: Discovered as an open-pollinated seedling somewhere in Canada. Selected and named by Sheridan Nurseries, Oakville, Canada, in 1966. Originally carried as (26).
Bibliography:
Dirr, M.A. *Manual of Woody Landscape Plants,* 4th Ed., 1990.
Flint, H.L. *Horticulture,* March 1987.
The Boxwood Bulletin, Vol. 7(1):1, July 1967/21(4):62, April 1982.
Known Locations: Arnold Arboretum, Secrest Arboretum, U.S. National Arboretum, State Arboretum of Virginia.
Additional Information:
Culture and Care: Prefers dappled shade but will tolerate siting in some direct sun.
Pests and Diseases: Resistance to leaf miner, psyllid and mites.
Available in the commercial nursery trade.

Buxus sinica variety *insularis* 'Tall Boy'

Size (25 yrs.):	Medium – 5 to 5^1/$_2$ feet high and 4^1/$_2$ to 5 feet wide.
Natural Form:	Conical – loose open habit.
Annual Growth:	Medium – 3 to 3^1/$_4$ inches in height and 2^1/$_2$ to 3 inches in width.
Leaf Color:	Medium green with yellow undertone.
Leaf Shape:	Lanceolate; obtuse tip; cuneate base.
Leaf Size:	Medium – ^{3}/$_4$ to ^{7}/$_8$ inch long and ^{1}/$_4$ to ^{5}/$_{16}$ inch wide.
Leaf Surface:	Glabrous – smooth.
Internodal Length:	Medium – ^{3}/$_8$ to ^{1}/$_2$ inch.
Flowering Habit:	Moderate flowering. Fruit set not observed.
Hardiness:	Zones 5 to 8.
Plant Use:	Specimen, grouping for background and area separations, hedgings.

6 ft.

Registration: B. Wagenknecht in *The Boxwood Bulletin*, Vol. 7(1):1, July 1967.
History: Discovered, selected and named by Sheridan Nurseries, Oakville, Ontario, Canada, from 100 boxwood seed received from Holland in 1946. Commercially released in 1960. Originally carried as (27).
Bibliography:
Dirr, M.A. *Manual of Woody Landscape Plants*, 4th Ed., 1990.
Known Locations: Secrest Arboretum, U.S. National Arboretum, State Arboretum of Virginia.
Additional Information:
Culture and Care: Prefers dappled shade but will tolerate siting in some direct sun.
Pests and Diseases: Resistance to leaf miner, psyllid and mites.
Not available in the commercial nursery trade.

Selected Species, Varieties, Cultivars

Buxus sinica variety *insularis* 'Tide Hill'

Size (25 yrs.):	Small – 2½ to 3 feet high and 4 to 4½ feet wide.
Natural Form:	Mounded – vase-shaped, compact.
Annual Growth:	Medium – 1 to 1½ inches in height and 2 to 2½ inches in width.
Leaf Color:	Medium yellow-green.
Leaf Shape:	Lanceolate; obtuse tip; cuneate base.
Leaf Size:	Medium – ½ to ¹¹/₁₆ inch long and ³/₁₆ to ⁵/₁₆ inch wide.
Leaf Surface:	Glabrous – smooth.
Internodal Length:	Long – ⅜ to ½ inch.
Flowering Habit:	Not observed.
Hardiness:	Zones 6 to 8.
Plant Use:	Specimen, grouping for background and area separations, edgings, hedgings.

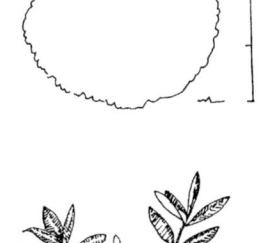

Registration: D. Wyman in *Arnoldia*, 17(11):64, 1957.
History: Discovered as an open-pollinated seedling on Samuel Everitt's Tide Hill estate, Huntington Bay, Long Island, N.Y., prior to 1936. Selected and named by Dr. B. Blackburn in 1954.
Bibliography:
Blackburn, B. *Popular Gardening*, 1964.
Bush-Brown, J. and L. *America's Garden Book,* Rev., New York Botanical Garden, 1980.
Dirr, M.A. *Manual of Woody Landscape Plants*, 4th Ed., 1990.
Wyman, D. *Wyman's Gardening Encyclopedia*, New Expanded 2nd Ed., 1986.
The Boxwood Bulletin, Vol. 25(3):68, Jan. 1986/27(3):65, Jan. 1988.
Known Locations: Arnold Arboretum, U.S. National Arboretum, State Arboretum of Virginia, Willowwood Arboretum.
Additional Information:
Culture and Care: Prefers dappled shade for the best performance. Keep it opened up, particularly in the top section, to provide ventilation and allow sunlight to penetrate and increase stem leaves to help support the feeder roots.
Pests and Diseases: Resistance to leaf miner, psyllid and mites.
Available in the commercial nursery trade.

Buxus sinica variety *insularis* 'Winter Beauty'

Size (25 yrs.):	Medium – 3 to $3^{1}/_{2}$ feet high and 5 to $5^{1}/_{2}$ feet wide.
Natural Form:	Mounded – loose open habit.
Annual Growth:	Medium – $1^{1}/_{2}$ to $1^{3}/_{4}$ inches in height and $2^{1}/_{2}$ to 3 inches in width.
Leaf Color:	Medium yellow-green.
Leaf Shape:	Lanceolate; obtuse tip; cuneate base.
Leaf Size:	Small – $^{1}/_{2}$ to $^{5}/_{8}$ inch long and $^{3}/_{16}$ to $^{5}/_{16}$ inch wide.
Leaf Surface:	Glabrous – smooth.
Internodal Length:	Medium – $^{1}/_{4}$ to $^{5}/_{16}$ inch.
Flowering Habit:	Floriferous; heavy fruiting.
Hardiness:	Zones 5(protected) to 8.
Plant Use:	Specimen, grouping for background and area separations, hedgings.

Registration: B. Wagenknecht in *The Boxwood Bulletin,* Vol. 7(1):1, July 1967.
History: Discovered, selected and named by Sheridan Nurseries, Oakville, Ontario, Canada, from 100 boxwood seed received from Holland in 1946. The plant was commercially released in 1967. Originally carried as (30).
Bibliography:
Dirr, M.A. *Manual of Woody Landscape Plants,* 4th Ed., 1990.
The Boxwood Bulletin, Vol. 7(1):1, July 1967/21(4):62, April 1982.
Known Locations: Arnold Arboretum, Secrest Arboretum, U.S. National Arboretum, State Arboretum of Virginia.
Additional Information:
Culture and Care: Prefers dappled shade but will tolerate siting in some direct sun.
Pests and Diseases: Resistance to leaf miner, psyllid and mites.
Available in the commercial nursery trade.

Buxus sinica variety *insularis* 'Wintergreen'

Size (25 yrs.):	Medium, variable – 2 to 4 feet high and 3 to 5 feet wide.
Natural Form:	Variable – mounded, somewhat compact to loose and open habit.
Annual Growth:	Medium – 1 to 2 inches in height and $1^1/_2$ to $2^1/_2$ inches in width.
Leaf Color:	Medium yellow-green.
Leaf Shape:	Obovate; obtuse tip; cuneate base.
Leaf Size:	Small – $^1/_2$ to $^5/_8$ inch long and $^1/_4$ to $^5/_{16}$ inch wide.
Leaf Surface:	Glabrous – smooth.
Internodal Length:	Long – $^3/_8$ to $^5/_8$ inch.
Flowering Habit:	Floriferous; heavy fruiting.
Hardiness:	Zones 5 to 8.
Plant Use:	Specimen, grouping for background and area separations, hedgings.

Registration: D. Wyman in *Arnoldia,* 23(5):88, Winter 1963.
History: Seed was imported from Manchuria by Scarff Nurseries, New Carlisle, Ohio, in the 1930s. Twenty-five seedlings were selected and the name Wintergreen given to all of them. Technically they are not the same clone but rather 25 separate clones; nonetheless, they were all lumped together and commercially released. Since each has some variations in its characteristics it becomes very difficult to describe the plant under one set of standards and various designations are beginning to turn up, i.e., 'Wintergreen,' 'Wintergreen 58,' and 'Wintergreen HNS.'
Bibliography:
Bush-Brown, J. and L. *America's Garden Book,* Rev., New York Botanical Garden, 1980.
Dirr, M.A. *Manual of Woody Landscape Plants,* 4th Ed., 1990.
Flint, H.L. *Horticulture,* March 1987.
Gamble, M.A. *Flower and Garden,* March 1988.
Krussman, G. *Manual of Cultivated Broad-leaved Trees and Shrubs,* Vol. I, A-D, 1986.
Wyman, D. *Wyman's Gardening Encyclopedia,* New Expanded 2nd Ed., 1986.
The Boxwood Bulletin, Vol. 3(1):1, July 1963/4(4):67, April 1965/16(1):14, July 1976/ 17(3):39, Jan. 1978/20(3):43,46, Jan. 1981/21(3):45, Jan. 1982/21(4):62, April 1982/27(3):65, Jan. 1988.
Known Locations: Arnold Arboretum, Missouri Botanical Garden, Secrest Arboretum, U.S. National Arboretum, State Arboretum of Virginia.
Additional Information:
Culture and Care: Will tolerate siting in some direct sun; occasionally suffers from winter bronzing.
Pests and Diseases: Resistance to leaf miner, psyllid and mites.
Available in the commercial nursery trade.

Buxus sinica variety *insularis* – Partial Descriptions

Cultivars with insufficient data for a reasonable plant description:

'Arnold Arboretum' – Catalog, Miami Nursery, Box 81, Tipp City, Ohio, 1985.

'Dansville' – Inventory, University of Wisconsin, Madison, Wis., 1986.

'Garden Variety' – Inventory, U.S. National Arboretum accession # (29693). Catalog, Kingsville Nursery, Kingsville, Md., 1961. *The Boxwood Bulletin*, Vol. 9(1):52, July 1969.

'Large Leaf Asiatic' – Catalog, John Vermeulen and Son, Neshanic Station, N.J.

'Staygreen' – Catalog, John Vermeulen and Son, Neshanic Station, N.J., 1961. Originated by Ernest Miller, Stonewall Gardens Nursery, Kent, Ohio. Inventory of Arnold Arboretum.

'Wintergreen 58' – Inventory Secrest Arboretum, 1982. *The Boxwood Bulletin*, Vol. 21(4):63, April 1982. "Two separate plantings have been growing for 8 years. They were outplanted as 14-inch container-grown plants. Boxwood set on site somewhat exposed to winds and winter sun were severely damaged with temperatures to -20° F."

'Wintergreen HNS' – Inventory Secrest Arboretum, 1982. *The Boxwood Bulletin*, Vol. 21(4):63, April 1982. "Two separate plantings have been growing for 8 years. They were outplanted as 12-inch container-grown plants. Boxwoods set on a site somewhat exposed to winds and winter sun had a few leaves killed at temperatures dropping to -20° F. even though they were completely covered by snow. Plants not completely covered by snow had all exposed wood killed up to 40 percent of the entire shrub. All plants suffered accumulative winter kill from three successive severe winters when temperatures were from -17° F. to -20° F. They have recovered somewhat during the three growing seasons since the period of severe winters. Plants are from 12 to 18 inches high with spreads of 17 to 20 inches and overall are only in fair condition. Foliage is bronze from early October freezes."

"Wintergreen HNS Boxwood set on a protected site only suffered twig kill up to 2 inches on exposed twigs above snow cover. There was accumulative damage following three severe winters in a row. Foliage greens up well during summers. Eight years after being set out these boxwood are from 22 to 36 inches high with spreads from 24 to 26 inches. Difference in height growth is largely the result of different degrees of winter kill. Crowns are open and irregular. Foliage is currently bronze from early October freezes. All plants have a heavy set of flower buds (Nov. 1981). Wintergreen HNS Box has been marginally hardy in the Arboretum (Secrest) on protected sites."

Buxus × 'Green Gem'

Size (25 yrs.):	Medium – 4½ to 5 feet high and 5 to 6 feet wide.
Natural Form:	Pyramidal, billowy with a slightly loose habit.
Annual Growth:	Medium – 2 to 2½ inches in height and 2½ to 3 inches in width.
Leaf Color:	Medium green with yellow undertones.
Leaf Shape:	Obovate to elliptic; obtuse to acute tip; cuneate base.
Leaf Size:	Medium – ¾ to ⅞ inch long and ⁵⁄₁₆ to ⅜ inch wide.
Leaf Surface:	Glabrous and smooth.
Internodal Length:	Medium – ⅜ to ⁵⁄₁₆ inch.
Flowering Habit:	Sparse flowering and not observed to set fruit.
Hardiness:	Zones 5 to 8.
Plant Use:	Specimen, foundation planting, grouping for background and area separations, hedgings.

Registration: B. Wagenknecht in *The Boxwood Bulletin,* Vol. 7(1):1, July 1967.
History: Originated, selected and named by Sheridan Nurseries, Oakville, Ontario, Canada.
Bibliography:
Flint, H.L. *Horticulture,* March 1987.
Gamble, M.A. *Flower and Garden,* March 1988.
The Boxwood Bulletin, Vol. 7(1):1, July 1967/29(3):49, Jan. 1990.
Known Locations: U.S. National Arboretum, State Arboretum of Virginia.
Additional Information:
Culture and Care: Demonstrates no special cultural requirements.
Pests and Diseases: Indicates an attraction for leaf miner and psyllid, with resistance to mites in the more humid climates; no serious diseases.
Available in the commercial nursery trade.

Buxus × 'Green Mound'

Size (25 yrs.):	Large – 6 to 6½ feet high and 5 to 5½ feet wide.
Natural Form:	Pyramidal, billowy and tending toward globular.
Annual Growth:	Medium – 3 to 3¼ inches in height and 2½ to 3 inches in width.
Leaf Color:	Medium green with yellow undertone.
Leaf Shape:	Elliptic; acute tip; cuneate base.
Leaf Size:	Medium – ¾ to ⅞ inch long and 5/16 to ⅜ inch wide.
Leaf Surface:	Glabrous and smooth.
Internodal Length:	Medium – ⅜ to 5/16 inch.
Flowering Habit:	Not observed.
Hardiness:	Zones 5 to 8.
Plant Use:	Specimen, foundation planting, grouping for background and area separations, hedgings.

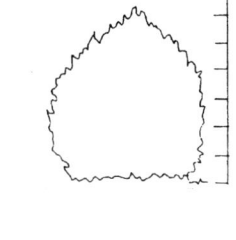

Registration: Not registered.
History: Originated, selected and named by Sheridan Nurseries, Oakville, Ontario, Canada.
Bibliography:
Dirr, M.A. *Manual of Woody Landscape Plants*, 4th Ed., 1990.
The Boxwood Bulletin, Vol. 29(3):49, Jan. 1990.
Known Locations: State Arboretum of Virginia.
Additional Information:
Culture and Care: Demonstrates no special cultural requirements.
Pests and Diseases: Indicates an attraction for leaf miner and psyllid, with resistance to mites in the more humid climates; no serious diseases.
Available in the commercial nursery trade.

Selected Species, Varieties, Cultivars

Buxus × 'Green Mountain'

Size (25 yrs.):	Medium – 5 to 5½ feet high and 6½ to 7 feet wide.
Natural Form:	Pyramidal, billowy and somewhat loose, open habit.
Annual Growth:	Medium – 2½ to 3 inches in height and 3½ to 4 inches in width.
Leaf Color:	Medium green.
Leaf Shape:	Elliptic; acute tip; cuneate base.
Leaf Size:	Medium – ¾ to ⅞ inch long and ³⁄₁₆ to ⁵⁄₁₆ inch wide.
Leaf Surface:	Glabrous and smooth.
Internodal Length:	Medium – ⅜ to ⁵⁄₁₆ inch.
Flowering Habit:	Sparse flowering and sparse fruiting.
Hardiness:	Zones 5 to 8.
Plant Use:	Specimen, foundation planting, grouping for background and area separations, hedgings.

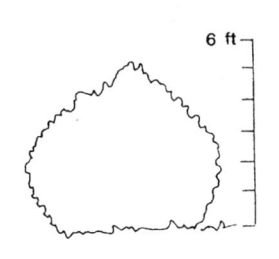

Registration: B. Wagenknecht in *The Boxwood Bulletin,* Vol. 7(1):1, July 1967.
History: Originated, selected and named by Sheridan Nurseries, Oakville, Ontario, Canada.
Bibliography:
Dirr, M.A. *Manual of Woody Landscape Plants*, 4th Ed., 1990.
Flint, H.L. *Horticulture,* March 1987.
The Boxwood Bulletin, Vol. 7(1):1, July 1967/29(3):49, Jan. 1990.
Known Locations: U.S. National Arboretum, State Arboretum of Virginia.
Additional Information:
Culture and Care: Demonstrates no special cultural requirements.
Pests and Diseases: Indicates an attraction for leaf miner and psyllid, with resistance to mites in the more humid climates; no serious diseases.
Available in the commercial nursery trade.

Buxus × 'Green Velvet'

Size (25 yrs.):	Medium – 4 to 5 feet high and 4 to 4½ feet wide.
Natural Form:	Pyramidal, somewhat billowy and loose, open habit.
Annual Growth:	Medium – 2 to 2½ inches in height and 2 to 2½ inches in width.
Leaf Color:	Medium green with yellow undertone.
Leaf Shape:	Obovate to elliptic; acute tip; cuneate base.
Leaf Size:	Medium – ¾ to ⅞ inch long and ⁵⁄₁₆ to ⅜ inch wide.
Leaf Surface:	Glabrous and smooth.
Internodal Length:	Medium – ⅜ to ⁵⁄₁₆ inch.
Flowering Habit:	Moderate flowering and moderate fruiting.
Hardiness:	Zones 5 to 8.
Plant Use:	Specimen, foundation planting, grouping for background and area separations, hedgings.

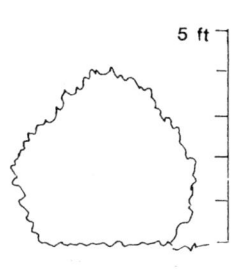

Registration: B. Wagenknecht in *The Boxwood Bulletin,* Vol. 7(1):1, July 1967.
History: Originated, selected and named by Sheridan Nurseries, Oakville, Ontario, Canada.
Bibliography:
Dirr, M.A. *Manual of Woody Landscape Plants,* 4th Ed., 1990.
Flint, H.L. *Horticulture,* March 1987.
Gamble, M.A. *Flower and Garden,* March 1988.
The Boxwood Society of the Midwest Bulletin, April 1986.
The Boxwood Bulletin, Vol. 7(1):1, July 1967/29(3):49, Jan. 1990.
Known Locations: Buzzards Bay Garden Club, Dixon Gallery and Garden, Missouri Botanical Garden, U.S. National Arboretum, State Arboretum of Virginia.
Additional Information:
Culture and Care: Demonstrates no special cultural requirements.
Pests and Diseases: Indicates an attraction for leaf miner and psyllid, with resistance to mites in the more humid climates; no serious diseases.
Available in the commercial nursery trade.

Plant Descriptions Summary

NAME	SIZE	PLANT (25 YEARS) HIGH	WIDE	NATURAL FORM	GROWTH RATE
B. austro-yunnanensis	LG	7	—	UNU	FA
B. bahamensis	LG	7	6	ARB	ME
B. balearica	LG	8	5	ARB	ME
B. bartletti	LG	15	—	ARB	FA
B. bodineri	ME	4	5	PYR	ME
B. colchica	ME	4	5	PYR	ME
B. hainanensis	LG	6	NOB	NOB	ME
B. harlandii	ME	4	5	VAS	ME
B. hebecarpa	LG	6	NOB	NOB	ME
B. henryi	LG	6	NOB	NOB	NOB
B. himalayensis	LG	8	6	PYR	FA
B. ichangensis	ME	4	NOB	NOB	ME
B. latistyla	LG	13	NOB	NOB	FA
B. linearifolia	SM	3	NOB	NOB	ME
B. megistophylla	LG	7	NOB	NOB	ME
B. microphylla	Widely variable				
B. microphylla variety japonica	ME	5	7	MON	FA
B. mollicula	LG	9	NOB	NOB	FA
B. myrica	LG	6	NOB	NOB	FA
B. pubiramea	LG	9	NOB	NOB	FA
B. rugulosa	ME	4	NOB	NOB	ME
B. sempervirens	Widely variable				
B. sinica	Widely variable				

PLANT SIZE
DW Dwarf
SM Small
ME Medium
LG Large

NATURAL FORM
ARB Arboreal
COL Columnar
CON Conical
MON Mound
OVA Ovate
PYR Pyramidal
SPH Spherical
VAS Vase
UNU Unusual

GROWTH RATE
SL Slow
ME Medium
FA Fast

HARDINESS
See page 48 for list of zones and map

LEAF COLOR
LIG Light green
MEG Medium green
DKG Dark green
LYG Light yellow-green
MYG Medium yellow-green
DYG Dark yellow-green
BIS Bi-colored, silver margin
BIG Bi-colored, gold margin

Selected Species, Varieties, and Cultivars

HARDI-NESS	LEAF COLOR	LEAF SHAPE	LEAF SIZE	LEAF SURFACE	INTER. LENGTH	FLOWER. HABIT	SEED PROD.
–	–	OBL	LG	GLAB	ME	MOD	MOD
9	MYG	ELL	LG	GLAB	LO	MOD	MOD
7	DKG	ELL	LG	GLAB	LO	FLO	HVY
9	MEG	ELL	LG	GLAB	ME	NOB	NOB
6	MEG	OBL	LG	GLOS	ME	MOD	MOD
6	MEG	ELL	ME	GLAB	ME	NOB	NOB
7	NOB	ELL	LG	GLAB	LO	MOD	MOD
6	MEG	OBO	LG	GLOS	ME	SPA	SPA
6	NOB	ELL	LG	GLAB	ME	MOD	MOD
6	NOB	ELL	ME	GLAB	ME	MOD	MOD
6	MEG	ELL	ME	GLAB	ME	SPA	SPA
6	NOB	OBL	ME	GLAB	ME	MOD	MOD
6	NOB	OVA	LG	GLAB	LO	MOD	MOD
6	NOB	ELL	LG	GLAB	SH	MOD	MOD
6	NOB	ELL	LG	GLOS	NOB	MOD	MOD
5	MYG	OVA	LG	GLOS	LO	FLO	HVY
6	NOB	OVA	ME	GLAB	ME	MOD	MOD
6	NOB	LAN	LG	GLAB	ME	MOD	MOD
6	NOB	LAN	LG	GLAB	NOB	NOB	NOB
6	NOB	LAN	ME	GLAB	ME	SPA	SPA

LEAF SHAPE
ELL Elliptic
LAN Lanceolate
OBL Oblanceolate
OBO Obovate
OVA Ovate
ROT Rotund
REV Revolute

LEAF SIZE
SM Small
ME Medium
LG Large

LEAF SURFACE
BULL Bullate
GLAB Glabrous
GLAC Glaucous
GLOS Glossy
MATT Matte

INTERNODAL LENGTH
SH Short
ME Medium
LO Long

FLOWERING HABIT/SEED PRODUCTION
FLO Floriferous
HVY Heavy seed set
MOD Moderate
SPA Sparse
NOB Not observed

Selected Species, Varieties, and Cultivars

Plant Descriptions Summary

NAME	PLANT (25 YEARS) SIZE	HIGH	WIDE	NATURAL FORM	GROWTH RATE
B. sinica variety insularis	ME	5	7	MON	SL
B. stenophylla	ME	5	NOB	NOB	ME
B. wallichiana	LG	8	6	PYR	FA
B. harlandii					
'Richard'	ME	5	5	UNU	ME
B. microphylla					
'Compacta'	DW	1	2	MON	SL
'Creepy'	DW	1	2	MON	SL
'Curly Locks'	ME	5	4	UNU	ME
'Grace Hendrick Phillips'	DW	1	3	MON	SL
'Green Pillow'	ME	3	4	MON	SL
'Helen Whiting'	ME	4	5	MON	SL
'Henry Hohman'	ME	5	6	UNU	ME
'Jim's True Spreader'	NOB	NOB	NOB	NOB	ME
'John Baldwin'	ME	5	4	PYR	ME
'Kingsville'	ME	3	3	MON	SL
'Locket'	ME	5	3	UNU	ME
'Miss Jones'	SM	3	4	MON	SL
'Quiet End'	ME	4	8	MON	ME
'Sport Compacta No. 1'	DW	1	2	MON	SL
'Sport Compacta No. 2'	DW	1	2	MON	SL

PLANT SIZE
DW Dwarf
SM Small
ME Medium
LG Large

NATURAL FORM
ARB Arboreal
COL Columnar
CON Conical
MON Mound

OVA Ovate
PYR Pyramidal
SPH Spherical
VAS Vase
UNU Unusual

GROWTH RATE
SL Slow
ME Medium
FA Fast

HARDINESS
See page 48 for list of zones and map

LEAF COLOR
LIG Light green
MEG Medium green
DKG Dark green
LYG Light yellow-green
MYG Medium yellow-green
DYG Dark yellow-green
BIS Bi-colored, silver margin
BIG Bi-colored, gold margin

Selected Species, Varieties, and Cultivars

HARDI-NESS	LEAF COLOR	LEAF SHAPE	LEAF SIZE	LEAF SURFACE	INTER. LENGTH	FLOWER. HABIT	SEED PROD.
5	LYG	OVA	SM	GLAB	LO	SPA	SPA
6	DKG	OBO	ME	GLAB	ME	MOD	MOD
7	DKG	ELL	LG	GLAB	ME	SPA	SPA
7	MEG	OBO	LG	GLOS	ME	NOB	NOB
5	LYG	OVA	SM	GLOS	SH	NOB	NOB
6	MEG	OBO	SM	GLOS	SH	NOB	NOB
5	LYG	OBO	SM	GLAB	SH	FLO	HVY
6	MYG	OBO	SM	GLOS	SH	NOB	NOB
5	DKG	OVA	SM	MATT	SH	NOB	NOB
6	LIG	OBO	SM	GLOS	SH	NOB	NOB
6	LYG	OBO	SM	GLAB	SH	SPA	SPA
6	MEG	OBO	ME	GLOS	ME	NOB	NOB
6	DKG	OBO	SM	GLAB	ME	MOD	MOD
6	MYG	LAN	LG	GLOS	SH	SPA	NOB
5	LIG	OBO	SM	GLAB	SH	MOD	MOD
6	MYG	LAN	SM	GLAB	ME	MOD	NOB
6	LYG	OBO	SM	GLAB	SH	NOB	NOB
5	MEG	LAN	SM	GLAB	SH	NOB	NOB
5	MEG	OBO	SM	GLAB	SH	NOB	NOB

LEAF SHAPE
ELL Elliptic
LAN Lanceolate
OBL Oblanceolate
OBO Obovate
OVA Ovate
ROT Rotund
REV Revolute

LEAF SIZE
SM Small
ME Medium
LG Large

LEAF SURFACE
BULL Bullate
GLAB Glabrous
GLAC Glaucous
GLOS Glossy
MATT Matte

INTERNODAL LENGTH
SH Short
ME Medium
LO Long

FLOWERING HABIT/SEED PRODUCTION
FLO Floriferous
HVY Heavy seed set
MOD Moderate
SPA Sparse
NOB Not observed

Selected Species, Varieties, and Cultivars

Plant Descriptions Summary

NAME	PLANT (25 YEARS) SIZE	HIGH	WIDE	NATURAL FORM	GROWTH RATE
'Sunlight'	ME	4	5	MON	ME
'Sunnyside'	LG	6	6	PYR	FA
'Winter Gem'	ME	5	4	PYR	ME
B. microphylla variety *japonica*					
'Green Beauty'	ME	4	4	MON	ME
'Morris Dwarf'	SM	2	3	MON	SL
'Morris Midget'	SM	2	3	MON	SL
'Nana Compacta'	DW	1	2	MON	SL
'National'	LG	9	8	PYR	FA
B. sempervirens					
'Abilene'	LG	7	8	PYR	FA
'Agram'	ME	5	6	PYR	ME
'Arborescens'	LG	11	8	ARB	FA
'Argenteo-variegata'	SM	2	4	MON	SL
'Aristocrat'	LG	10	6	PYR	FA
'Asheville'	ME	5	3	PYR	ME
'Aurea Pendula'	LG	8	10	UNU	FA
'Aureo-variegata'	LG	7	7	PYR	FA
'Belleville'	LG	7	8	PYR	ME
'Blauer Heinz'	DW	2	2	MON	SL
'Bullata'	LG	8	10	PYR	FA

PLANT SIZE
DW Dwarf
SM Small
ME Medium
LG Large

NATURAL FORM
ARB Arboreal
COL Columnar
CON Conical
MON Mound
OVA Ovate
PYR Pyramidal
SPH Spherical
VAS Vase
UNU Unusual

GROWTH RATE
SL Slow
ME Medium
FA Fast

HARDINESS
See page 48 for list of zones and map

LEAF COLOR
LIG Light green
MEG Medium green
DKG Dark green
LYG Light yellow-green
MYG Medium yellow-green
DYG Dark yellow-green
BIS Bi-colored, silver margin
BIG Bi-colored, gold margin

Selected Species, Varieties, and Cultivars

HARDI-NESS	LEAF COLOR	LEAF SHAPE	LEAF SIZE	LEAF SURFACE	INTER. LENGTH	FLOWER. HABIT	SEED PROD.
5	LYG	OBO	SM	GLOS	ME	NOB	NOB
5	MYG	ROT	LG	GLOS	ME	MOD	MOD
5	MYG	OVA	ME	GLOS	ME	NOB	NOB
6	DKG	LAN	ME	GLOS	ME	NOB	NOB
5	MYG	OBO	SM	GLAB	SH	NOB	NOB
5	MYG	OBO	SM	GLAB	SH	NOB	NOB
5	MEG	OVA	SM	GLOS	SH	NOB	NOB
6	DKG	ROT	LG	GLOS	LO	SPA	SPA
5	MEG	LAN	LG	GLAB	LO	NOB	NOB
5	MEG	LAN	ME	GLAB	ME	SPA	SPA
5	MEG	OVA	LG	GLAB	LO	SPA	SPA
6	BIS	LAN	LG	BULL	ME	NOB	NOB
5	MEG	LAN	ME	GLOS	LO	NOB	NOB
5	DKG	LAN	ME	GLAB	ME	NOB	NOB
6	BIG	OVA	ME	GLAB	LO	NOB	NOB
6	BIG	OVA	LG	GLAB	LO	SPA	SPA
5	DKG	ELL	ME	GLAB	ME	MOD	MOD
5	DKG	OBO	SM	GLAB	SH	NOB	NOB
6	DKG	LAN	LG	BULL	LO	FLO	HVY

LEAF SHAPE
ELL Elliptic
LAN Lanceolate
OBL Oblanceolate
OBO Obovate
OVA Ovate
ROT Rotund
REV Revolute

LEAF SIZE
SM Small
ME Medium
LG Large

LEAF SURFACE
BULL Bullate
GLAB Glabrous
GLAC Glaucous
GLOS Glossy
MATT Matte

INTERNODAL LENGTH
SH Short
ME Medium
LO Long

FLOWERING HABIT/SEED PRODUCTION
FLO Floriferous
HVY Heavy seed set
MOD Moderate
SPA Sparse
NOB Not observed

Selected Species, Varieties, and Cultivars

Plant Descriptions Summary

NAME	PLANT (25 YEARS) SIZE	PLANT (25 YEARS) HIGH	PLANT (25 YEARS) WIDE	NATURAL FORM	GROWTH RATE
'Butterworth'	LG	7	6	OVA	ME
'Clembrook'	ME	4	3	PYR	ME
'Cliffside'	LG	9	3	CON	FA
'Decussata'	LG	8	5	PYR	FA
'Dee Runk'	LG	10	2	COL	FA
'Denmark'	LG	8	8	PYR	FA
'Edgar Anderson'	ME	4	4	PYR	ME
'Elegantissima'	LG	7	6	PYR	FA
'Fastigiata'	LG	9	2	CON	FA
'Flora Place'	LG	6	4	PYR	ME
'Fortunei Rotundifolia'	LG	6	7	PYR	ME
'Glauca'	LG	6	5	PYR	ME
'Graham Blandy'	LG	10	1	COL	FA
'Handsworthiensis'	LG	8	8	PYR	FA
'Handsworthii'	LG	7	6	PYR	FA
'Hardwickensis'	ME	4	3	UNU	ME
'Heinrich Bruns'	ME	4	4	UNU	ME
'Henry Shaw'	ME	5	4	PYR	ME
'Hermann von Schrenk'	LG	7	9	PYR	ME
'Holland'	ME	5	4	CON	ME
'Hood'	ME	3	3	PYR	SL
'Inglis'	LG	9	8	PYR	FA
'Ipek'	LG	10	6	PYR	FA

PLANT SIZE
DW Dwarf
SM Small
ME Medium
LG Large

NATURAL FORM
ARB Arboreal
COL Columnar
CON Conical
MON Mound
OVA Ovate
PYR Pyramidal
SPH Spherical
VAS Vase
UNU Unusual

GROWTH RATE
SL Slow
ME Medium
FA Fast

HARDINESS
See page 48 for list of zones and map

LEAF COLOR
LIG Light green
MEG Medium green
DKG Dark green
LYG Light yellow-green
MYG Medium yellow-green
DYG Dark yellow-green
BIS Bi-colored, silver margin
BIG Bi-colored, gold margin

Selected Species, Varieties, and Cultivars

HARDI-NESS	LEAF COLOR	LEAF SHAPE	LEAF SIZE	LEAF SURFACE	INTER. LENGTH	FLOWER. HABIT	SEED PROD.
6	MEG	LAN	LG	GLOS	LO	SPA	NOB
5	DKG	LAN	ME	GLAB	ME	NOB	NOB
6	MEG	LAN	ME	GLOS	ME	NOB	NOB
6	DKG	LAN	LG	GLAB	LO	MOD	MOD
6	MEG	ELL	LG	GLAB	ME	MOD	MOD
5	MYG	ROT	LG	GLOS	ME	SPA	SPA
5	MEG	ELL	SM	GLAB	SH	NOB	NOB
6	BIG	LAN	ME	GLAB	LO	NOB	NOB
6	DKG	LAN	LG	GLOS	LO	SPA	SPA
5	DKG	LAN	LG	GLAB	LO	SPA	SPA
6	MYG	ROT	ME	GLOS	LO	MOD	MOD
6	DKG	LAN	LG	GLAC	LO	FLO	HVY
6	MEG	LAN	LG	GLOS	ME	NOB	NOB
6	DKG	LAN	LG	GLOS	ME	FLO	HVY
6	DKG	LAN	LG	MATT	SH	MOD	MOD
6	DKG	ROT	SM	GLAB	ME	NOB	NOB
6	DKG	OVA	ME	GLAB	ME	NOB	NOB
5	MEG	LAN	ME	GLAB	ME	MOD	SPA
5	MEG	ELL	ME	GLAB	SH	NOB	NOB
6	MEG	OVA	ME	GLAB	ME	NOB	NOB
5	DKG	LAN	ME	GLAB	ME	NOB	NOB
5	MEG	ELL	SM	GLAB	SH	FLO	HVY
5	DKG	LAN	LG	GLOS	ME	SPA	SPA

LEAF SHAPE
ELL Elliptic
LAN Lanceolate
OBL Oblanceolate
OBO Obovate
OVA Ovate
ROT Rotund
REV Revolute

LEAF SIZE
SM Small
ME Medium
LG Large

LEAF SURFACE
BULL Bullate
GLAB Glabrous
GLAC Glaucous
GLOS Glossy
MATT Matte

INTERNODAL LENGTH
SH Short
ME Medium
LO Long

FLOWERING HABIT/SEED PRODUCTION
FLO Floriferous
HVY Heavy seed set
MOD Moderate
SPA Sparse
NOB Not observed

Selected Species, Varieties, and Cultivars

Plant Descriptions Summary

NAME	PLANT (25 YEARS) SIZE	HIGH	WIDE	NATURAL FORM	GROWTH RATE
'Joe Gable'	ME	4	6	PYR	ME
'Joy'	LG	9	7	PYR	FA
'Krossi-livonia'	LG	12	11	PYR	FA
'Latifolia'	LG	6	7	PYR	FA
'Latifolia Marginata'	ME	5	6	MON	ME
'Liberty'	ME	4	3	OVA	ME
'Macrophylla'	LG	7	7	PYR	FA
'Maculata'	LG	7	7	PYR	FA
'Mary Gamble'	SM	3	3	SPH	SL
'Memorial'	ME	4	2	COL	ME
'Myosotidifolia'	LG	8	9	PYR	FA
'Myrtifolia'	ME	3	2	PYR	ME
'Natchez'	DW	2	3	MON	SL
'Newport Blue'	ME	4	5	PYR	ME
'Nish'	ME	3	4	PYR	ME
'Northern Find'	LG	6	9	MON	FA
'Northland'	LG	6	7	PYR	ME
'Notata'	LG	8	11	MON	FA
'Pendula'	LG	8	6	UNU	FA
'Ponteyi'	LG	6	7	PYR	ME
'Prostrata'	ME	5	7	UNU	ME
'Pullman'	LG	6	6	CON	ME
'Pyramidalis'	LG	8	4	CON	FA

PLANT SIZE
DW Dwarf
SM Small
ME Medium
LG Large

NATURAL FORM
ARB Arboreal
COL Columnar
CON Conical
MON Mound

OVA Ovate
PYR Pyramidal
SPH Spherical
VAS Vase
UNU Unusual

GROWTH RATE
SL Slow
ME Medium
FA Fast

HARDINESS
See page 48 for list of zones and map

LEAF COLOR
LIG Light green
MEG Medium green
DKG Dark green
LYG Light yellow-green
MYG Medium yellow-green
DYG Dark yellow-green
BIS Bi-colored, silver margin
BIG Bi-colored, gold margin

Selected Species, Varieties, and Cultivars

HARDI-NESS	LEAF COLOR	LEAF SHAPE	LEAF SIZE	LEAF SURFACE	INTER. LENGTH	FLOWER. HABIT	SEED PROD.
6	MEG	ELL	ME	GLAB	ME	NOB	NOB
5	MEG	LAN	ME	GLAB	ME	NOB	NOB
6	MEG	LAN	ME	GLAB	LO	FLO	HVY
6	DKG	LAN	ME	GLAB	ME	MOD	MOD
6	BIS	LAN	LG	BULL	LO	NOB	NOB
5	DKG	LAN	LG	GLAB	LO	MOD	MOD
6	DKG	OVA	ME	GLOS	ME	FLO	HVY
6	DKG	OVA	ME	GLOS	ME	FLO	HVY
5	MEG	LAN	SM	GLAB	SH	NOB	NOB
6	MEG	LAN	LG	GLAB	ME	NOB	NOB
6	MEG	LAN	LG	GLAB	LO	FLO	HVY
5	MEG	ELL	ME	GLOS	ME	NOB	NOB
5	MEG	ELL	ME	GLAB	ME	NOB	NOB
6	MEG	LAN	ME	GLOS	ME	NOB	NOB
5	DKG	ELL	ME	GLAB	ME	NOB	NOB
5	DKG	LAN	LG	GLAB	LO	MOD	MOD
5	DKG	LAN	ME	GLAB	ME	FLO	HVY
6	BIG	LAN	ME	GLAB	ME	NOB	NOB
5	MEG	LAN	LG	GLAB	LO	FLO	HVY
6	DKG	LAN	LG	GLAB	LO	FLO	HVY
6	DKG	ELL	LG	GLAB	LO	FLO	HVY
5	DKG	ELL	ME	GLAB	ME	NOB	NOB
6	MEG	ELL	LG	GLAB	ME	FLO	HVY

LEAF SHAPE
ELL Elliptic
LAN Lanceolate
OBL Oblanceolate
OBO Obovate
OVA Ovate
ROT Rotund
REV Revolute

LEAF SIZE
SM Small
ME Medium
LG Large

LEAF SURFACE
BULL Bullate
GLAB Glabrous
GLAC Glaucous
GLOS Glossy
MATT Matte

INTERNODAL LENGTH
SH Short
ME Medium
LO Long

FLOWERING HABIT/SEED PRODUCTION
FLO Floriferous
HVY Heavy seed set
MOD Moderate
SPA Sparse
NOB Not observed

Selected Species, Varieties, and Cultivars

Plant Descriptions Summary

NAME	PLANT (25 YEARS) SIZE	HIGH	WIDE	NATURAL FORM	GROWTH RATE
'Pyramidalis Hardwickensis'	LG	10	3	CON	FA
'Rochester'	ME	3	4	PYR	SL
'Rotundifolia'	LG	8	10	MON	FA
'Salicifolia'	LG	6	7	MON	FA
'Salicifolia Elata'	LG	6	11	MON	ME
'Ste. Genevieve'	LG	6	8	PYR	FA
'Suffruticosa'	DW	2	2	SPH	SL
'Tennessee'	ME	4	5	PYR	ME
'Undulifolia'	LG	10	8	ARB	FA
'Vardar Valley'	ME	3	8	MON	SL
'Welleri'	ME	3	4	PYR	ME
'West Ridgeway'	SM	2	2	SPH	SL
'Woodland'	ME	5	6	PYR	ME
'Yorktown'	ME	5	5	PYR	ME
'Zehtung'	ME	4	3	PYR	ME
B. sinica variety *insularis*					
'Justin Brouwers'	SM	2	3	MON	SL
'Nana'	SM	2	3	MON	SL
'Pincushion'	SM	3	4	MON	ME
'Tall Boy'	ME	5	4	CON	ME
'Tide Hill'	SM	2	4	MON	ME
'Winter Beauty'	ME	3	5	MON	ME

PLANT SIZE
DW Dwarf
SM Small
ME Medium
LG Large

NATURAL FORM
ARB Arboreal
COL Columnar
CON Conical
MON Mound

OVA Ovate
PYR Pyramidal
SPH Spherical
VAS Vase
UNU Unusual

GROWTH RATE
SL Slow
ME Medium
FA Fast

HARDINESS
See page 48 for list of zones and map

LEAF COLOR
LIG Light green
MEG Medium green
DKG Dark green
LYG Light yellow-green

MYG Medium yellow-green
DYG Dark yellow-green
BIS Bi-colored, silver margin
BIG Bi-colored, gold margin

Selected Species, Varieties, and Cultivars

HARDI-NESS	LEAF COLOR	LEAF SHAPE	LEAF SIZE	LEAF SURFACE	INTER. LENGTH	FLOWER. HABIT	SEED PROD.
5	DKG	OBO	LG	GLAB	LO	NOB	NOB
5	MEG	ELL	ME	GLAB	ME	NOB	NOB
6	MEG	ROT	LG	GLOS	LO	FLO	HVY
6	DYG	ELL	ME	GLAB	LO	FLO	HVY
6	MEG	ELL	LG	GLOS	LO	MOD	MOD
5	MEG	LAN	ME	GLAB	SH	MOD	MOD
6	MEG	OBO	ME	GLOS	ME	SPA	SPA
5	MEG	ELL	ME	GLOS	ME	NOB	NOB
6	DKG	LAN	LG	GLAB	ME	FLO	HVY
5	MEG	ROT	LG	GLAB	LO	SPA	SPA
5	MEG	ELL	ME	GLAB	ME	SPA	SPA
6	MEG	OBO	SM	GLAB	ME	NOB	NOB
6	DKG	LAN	LG	GLAB	ME	NOB	NOB
6	MYG	LAN	LG	GLOS	ME	NOB	NOB
5	MEG	LAN	ME	GLAB	ME	NOB	NOB
6	DKG	LAN	SM	GLAB	SH	NOB	NOB
6	MYG	LAN	SM	GLAB	ME	NOB	NOB
5	MEG	ROT	SM	GLAB	ME	MOD	MOD
5	MEG	LAN	ME	GLAB	ME	MOD	NOB
6	MYG	LAN	ME	GLAB	LO	NOB	NOB
5	MYG	LAN	SM	GLAB	ME	FLO	HVY

LEAF SHAPE
ELL Elliptic
LAN Lanceolate
OBL Oblanceolate
OBO Obovate
OVA Ovate
ROT Rotund
REV Revolute

LEAF SIZE
SM Small
ME Medium
LG Large

LEAF SURFACE
BULL Bullate
GLAB Glabrous
GLAC Glaucous
GLOS Glossy
MATT Matte

INTERNODAL LENGTH
SH Short
ME Medium
LO Long

FLOWERING HABIT/SEED PRODUCTION
FLO Floriferous
HVY Heavy seed set
MOD Moderate
SPA Sparse
NOB Not observed

Selected Species, Varieties, and Cultivars

Plant Descriptions Summary

NAME	PLANT (25 YEARS) SIZE	HIGH	WIDE	NATURAL FORM	GROWTH RATE
'Wintergreen'	ME	3	4	MON	ME
B. x					
'Green Gem'	ME	4	5	PYR	ME
'Green Mound'	LG	6	5	PYR	ME
'Green Mountain'	ME	5	6	PYR	ME
'Green Velvet'	ME	4	4	PYR	ME

PLANT SIZE
DW Dwarf
SM Small
ME Medium
LG Large

NATURAL FORM
ARB Arboreal
COL Columnar
CON Conical
MON Mound

OVA Ovate
PYR Pyramidal
SPH Spherical
VAS Vase
UNU Unusual

GROWTH RATE
SL Slow
ME Medium
FA Fast

HARDINESS
See page 48 for list of zones and map

LEAF COLOR
LIG Light green
MEG Medium green
DKG Dark green
LYG Light yellow-green
MYG Medium yellow-green
DYG Dark yellow-green
BIS Bi-colored, silver margin
BIG Bi-colored, gold margin

214 *Selected Species, Varieties, and Cultivars*

HARDI-NESS	LEAF COLOR	LEAF SHAPE	LEAF SIZE	LEAF SURFACE	INTER. LENGTH	FLOWER. HABIT	SEED PROD.
5	MYG	OBO	SM	GLAB	LO	FLO	HVY
5	MEG	OBO	ME	GLAB	ME	SPA	NOB
5	MEG	ELL	ME	GLAB	ME	NOB	NOB
5	MEG	ELL	ME	GLAB	ME	SPA	SPA
5	MEG	OBO	ME	GLAB	ME	MOD	MOD

LEAF SHAPE
ELL Elliptic
LAN Lanceolate
OBL Oblanceolate
OBO Obovate
OVA Ovate
ROT Rotund
REV Revolute

LEAF SIZE
SM Small
ME Medium
LG Large

LEAF SURFACE
BULL Bullate
GLAB Glabrous
GLAC Glaucous
GLOS Glossy
MATT Matte

INTERNODAL LENGTH
SH Short
ME Medium
LO Long

FLOWERING HABIT/SEED PRODUCTION
FLO Floriferous
HVY Heavy seed set
MOD Moderate
SPA Sparse
NOB Not observed

Selected Species, Varieties, and Cultivars

GLOSSARY

abortive: defective, barren, not developed.

adventitious: used to describe roots, buds and shoots from tissue of a different origin, e.g. roots formed on a stem cutting.

alternate: an arrangement of leaves or other parts not opposite or whorled; parts situated at a node, as leaves on a stem.

angustifolius: narrow-leaved.

apetalous: without petals.

arboreal: tree-like or pertaining to trees.

aureo: golden.

axillary: occurring in the angle formed by the leaf stalk and the branch.

bisexual: stamens and pistils present in one flower.

bottom heat: heat applied to plants from below, employed usually in a frame or greenhouse for propagating plants from cuttings or seed. Some boxwoods require this bottom heat to root or sprout readily.

bronzing: turning a metallic bronze or coppery color, especially of foliage during or shortly after winter.

bud: a structure of embryonic tissues, which will become a leaf, a flower, or both, or a new shoot.

bullate: with the surface appearing as if blistered or puckered.

capsule: in seed plants, a dry fruit composed of two or more chambers (valves) that break open to release the seeds.

chlorosis: the fading of the bright green color in leaves to pale green or yellow. Caused by a variety of diseases and by deficiencies of minerals such as potassium or iron.

clone: a group of plants derived vegetatively from one parent plant identical to each other and to the parent.

cultivar: a cultivated variety, an assemblage of cultivated plants that are distinguished by any character (morphological, cytological or others) and which, when reproduced (sexually or asexually), retain their distinguishing characteristics.

cuneate: wedge-shaped with essentially straight sides, the structure attached at the narrow end.

dehiscent: bursting open, the sides or segments of the splitting organ usually termed valves.

dioecious: having staminate and pistillate flowers borne on different individuals.

dormant: in a restive or non-vegetative state, especially a winter condition.

entire: having a margin without teeth or crenations. A leaf edge completely without indentation.

established: growing and reproducing without cultivation.

fertilizer: material rich in elements vital to plant growth added to the soil to increase its fertility. Fertilizers may be organic materials or inorganic substances (chemical fertilizers). The main elements supplied by fertilizers are nitrogen, phosphorus, potassium, magnesium, sulphur and the trace elements (those needed in minute quantities).

friable: easily crumbled, pulverized or reduced to powder.

globose: globular.

hardiness: used in temperate and cold climates to indicate the degree to which a plant is susceptible to frost damage. A hardy plant can live year-round without protection, a half-hardy or tender one needs protection during the coldest months of the year. In hot countries, "hardiness" may refer to a plant's resistance to drought.

heel: a small piece of two-year-old wood taken with a cutting of one-year-old wood for a certain method of propagation.

internode: the part of a plant stem between the joints (nodes), where leaf buds or leaves arise. Cuttings taken between joints are known as internodal cuttings.

leaching: the loss of nutrients in the soil when rain dissolves them and carries them beyond the reach of roots. It is a cause of infertility of light soils in high rainfall areas and a reason why fertilizers should be applied in small amounts and often rather than in large doses.

loam: soil consisting of clay, sand and organic materials in varying proportions.

longifolius: long-leaved.

macro: large or long.

maculatus: spotted.

mature: a later phase of growth characterized by flowering, fruiting and a reduced rate of size increase.

microclimate: the climate of a small area, which differs from that of the general surrounding region because of special conditions.

monoecious: a species with unisexual flowers; but having flowers of both sexes present on the same plant. See dioecious.

mutation: the change in the constitution of hereditary material (the gene) that gives rise to a new form, a mutant or sport. May occur spontaneously or be induced by agents such as chemicals or X-rays and may affect the whole plant or only part of it. See also sport.

nematode: a group of tiny worm-like creatures that may or may not damage plant roots.

node: a joint on a stem, represented by point of origin of a leaf or bud.

opposite: two at a node, as leaves.

pH: a measure of acidity and alkalinity used in reference to soil. pH values range from 0 to 14. A pH less than 7 indicates acidity, one more than 7 alkalinity. pH 7 is neutral.

pistillate: related to the female organs of a flower.

rooting hormones: preparations of hormones (growth-regulating substances) usually in powder form, into which cuttings are dipped to encourage root formation. Different preparations are available to treat soft, semi-woody, and woody cuttings.

sempervirens: evergreen.

species: a natural group of plants composed of similar individuals that can produce similar offspring; usually including several minor variations.

sport: an individual exhibiting, in whole or in part, a sudden spontaneous deviation from type beyond the normal limits of individual variation, usually as a result of a somatic mutation.

staminate: related to the male part of a flower.

subglobose: imperfectly or nearly globose.

taxonomy: the area of botany dealing with the classifying and naming of plants.

terminal: occurring at the tip of a branch.

unisexual: a flower with either stamens or pistil; of a single sex.

valve: one of the sections into which an opening fruit capsule separates.

variety: term used loosely by gardeners for any distinct form of a species or hybrid. Strictly speaking, all varieties in cultivation should be known as cultivars. Variety really means a true-breeding variant of a species in the wild.

x: indicates a hybrid; the offspring of two different species, varieties, cultivars.

SELECTED REFERENCES

Alphin, Thomas H. A descriptive study of varietal forms in *Buxus*. *American Journal of Botany,* 27:349-357, 1940.

Batdorf, L.R. *International Registration List of Cultivated Buxus L.* Boyce, Va.: The American Boxwood Society, 1987.

Brooklyn Botanic Gardens. *Plants and Garden Handbooks.* Brooklyn, N.Y.: Vol. 12, no. 3, 1967; Vol. 29, no. 3, 1973.

Burford, G.G. *Boxwood in the Landscape.* Boyce, Va.: Orland E. White Arboretum. 1993.

Bush-Brown. *America's Garden Book, Rev.* Bronx, N.Y.: New York Botanical Garden, 1980.

Dallimore, W.H. *Holly, Yew and Box.* London and New York: John Lane, 1908.

Dirr, M.A. *Manual of Woody Landscape Plants,* 4th Ed. Champaign, Ill.: Stipes Publishing Co., 1990.

Dudley, T.R. and G.K. Eisenbeiss. "Registration and Documentation of Cultivar Names." *The Boxwood Bulletin.* 11:12-14, 1971.

Everett, T.H. *The New York Botanical Garden Illustrated Encyclopedia of Horticulture.* New York: Garland STP Press, 1981.

Hartmann, Kester. *Plant Propagation; Principles and Practices,* 3rd Ed. Englewood Cliffs, N.J.: Prentice-Hall Inc., 1975.

Hottes, A.C. *The Book of Shrubs.* New York: A.T. DeLamare Co. Inc., 1931.

Johnson, W.T. and H.H. Lyon. *Insects That Feed on Trees and Shrubs,* 2nd Ed. Ithaca, N.Y.: Comstock Publishing Association Division of Cornell University Press, 1988.

Krussman, G. *Manual of Cultivated Broad-leaved Trees and Shrubs.* Vol. I, A-D. Portland, Ore.: Timber Press, 1986.

Larson, P.D. *Guide to the Natural Forms of Boxwood.* Boyce, Va.: Orland E. White Arboretum, the State Arboretum of Virginia, 1989.

Lewis, Albert Addison. *Boxwood Gardens, Old and New.* Richmond, Va.: The William Byrd Press Inc., 1924.

Loudon, J.C. *Arboretum et Fruticum Britannicum III,* 2nd Ed. London. pp. 1334-1335. 1844.

Mathes, M.C. *Collection of Woody Species.* Williamsburg, Va.: The College of William and Mary, 1980.

McCarty, Clara S. *The Story of Boxwood.* Richmond, Va.: The Dietz Press Inc., 1950.

Muenscher, W.C. *Poisonous Plants of the United States.* Revised edition. New York: Collier Books, 1975.

Rehder, Alfred. *Buxaceae,* in *Manual of Cultivated Trees and Shrubs.* Macmillan Co. pp. 536-538. 1960.

Sinclair, W.A., H.H. Lyon and W.T. Johnson. *Diseases of Trees and Shrubs.* Ithaca, N.Y.: Comstock Publishing Association Division of Cornell University Press, 1987.

Smith, A.G., Jr. *The Boxwood at Stratford Hall.* Stratford, Va.: The Robert E. Lee Memorial Foundation, 1965.

Staples, M.J. C. "A History of Box in the British Isles Part I." *The Boxwood Bulletin.* 10:18-23, 1970.

Stoetcel, M.B. *Common Names of Insects and Related Organisms.* Lanham, Md.: Entomological Society of America, 1989.

Wyman, D. *Wyman's Gardening Encyclopedia,* New Expanded 2nd Ed. New York: MacMillan Co., 1986.

Periodicals

The American Boxwood Society, *The Boxwood Bulletin.* Issued quarterly since 1961. Boyce, Va.

The Boxwood Society of the Midwest, *Bulletin.* Issued bi-annually since 1976. St. Louis, Mo.

REGISTRATION LIST

The author has taken the liberty to include the individual amendments to this document since it was first published in April 1987.

INTERNATIONAL REGISTRATION LIST OF CULTIVATED *BUXUS* L.

Lynn R. Batdorf

Curator, U.S. National Arboretum

International Registration Authority for Cultivated *Buxus* L.

Several considerations led to the preparation of this list. Primarily, with numerous additions to the original registration list published in 1965, a consolidation seemed desirable. Additionally, there was a need to publish a list of *approved common names* used for boxwoods. Finally, to prevent continued use of invalid cultivar names, those improperly or mistakenly cited in literature, it seemed appropriate to remove these names from the current registration list and place them in a separate list together with their proper synonyms.

Accordingly, the reader will find not one list but three lists following. The first list contains the approved botanical and common names of plants in American commerce and use; it is drawn from a directory of Standardized Plant Names published by the American Joint Committee on Horticultural Nomenclature. The second list is a current registration list of valid *Buxus* L. cultivars registered by the previous and present international registration authorities for cultivated *Buxus* L. The third list contains invalid names — together with their valid synonyms — of cultivars previously included in the Registration List of Cultivated *Buxus* L.

In all three lists, valid names are given in boldface type and invalid names in lightface type. In the current registration list (the second list below) the cultivar name is followed by the first valid citation found in literature.

I. Standardized Plant Names

Botanical Name	Common Name
Buxus balearica	Spanish Boxwood
Buxus harlandii	Harlands Boxwood
Buxus microphylla	Littleleaf Boxwood
Buxus microphylla var. *japonica*	Japanese Littleleaf Boxwood
Buxus sempervirens	Common Boxwood
Buxus sempervirens 'Angustifolia'	Willow Boxwood
Buxus sempervirens 'Arborescens'	Truetree Boxwood
Buxus sempervirens 'Argenteo-variegata'	Silver Boxwood
Buxus sempervirens 'Aurea Pendula'	Golden Weeping Boxwood
Buxus sempervirens 'Aureo-variegata'	Golden Boxwood
Buxus sempervirens 'Elegans'	Variegated Olive Boxwood
Buxus sempervirens 'Handsworthii'	Handsworth Boxwood
Buxus sempervirens 'Marginata'	Goldedge Boxwood
Buxus sempervirens 'Myosotidifolia'	Forgetmenotleaf Boxwood
Buxus sempervirens 'Myrtifolia'	Myrtleleaf Boxwood
Buxus sempervirens 'Pendula'	Weeping Boxwood
Buxus sempervirens 'Prostrata'	Prostrate Boxwood
Buxus sempervirens 'Pyramidalis'	Pyramid Boxwood
Buxus sempervirens 'Rosmarinifolia'	Rosemary Boxwood
Buxus sempervirens 'Rotundifolia'	Roundleaf Boxwood
Buxus sempervirens 'Suffruticosa'	Truedwarf Boxwood
Buxus sinica var. *insularis*	Chinese Littleleaf Boxwood
Buxus wallichiana	Wallichian Boxwood

II. Registration List of Cultivated *Buxus* L.

Buxus balearica Lam. in *Encyc. Meth. Bot.* 1:511.1785.
'Marginata' P. Corbelli in *Dizionario di Floricultura* 1:231.1873.

Buxus harlandii Hance in *Journal of Linnean Society* 13:123.1873.
'Richard' J. Baldwin in *The Boxwood Bulletin* 2(4):44.1963.

Buxus microphylla P.F. Siebold & J.C. Zuccarini in *Abhandl. Math. Phys. Konigl. Akad. Wissensch. Munch.* 4(2):142.1845.
'Compacta' D. Wyman in *American Nurseryman* 107(7):50.1963.
'Curly Locks' D. Wyman in *American Nurseryman* 107(7):50.1963.
'Grace Hendrick Phillips' H. Hohman in *The Boxwood Bulletin* 7(1):1.1967.
'Green Pillow' O'Connor in *Baileya* 1:114.1963.
'Green Sofa' J. Baldwin in *The Boxwood Bulletin* 15(3):42.1976.
'Helen Whiting' J. Baldwin in *The Boxwood Bulletin* 15(3):41-42.1976.
'John Baldwin' P.D. Larson in *The Boxwood Bulletin* 28(2):27.1988.
'Locket' J. Baldwin in *The Boxwood Bulletin* 15(3):41.1976 - 16(1):10-11.1976.
'Quiet End' L.R. Batdorf in *The Boxwood Bulletin* 31(3):52.1992.

Buxus microphylla var. *japonica* (J. Mueller, Arg.) Rehder and Wilson in Sargent, *Plantae Wilsonianae* 2(1):168.1914.
'Alba' Catalog, Andorra Nurseries, Chestnut Hill, Philadelphia, Pa. 1908.
'Angustifolia' L.H. Bailey in *Hortus* 105.1930.
'Argentea' Beissner, Schelle and Zabel, *Handbuch der Laubholz-Benennung* 283.1903.
'Aurea' Catalog, Charles Dietriche, Angers, France. 1892.
'Fortunei' Catalog, Andorra Nurseries, Chestnut Hill, Philadelphia, Pa. 1908.
'Japanese Globe' Plant List, K. Howell, 4100 E. Sprague St., Spokane, Wash. 1958.
'Latifolia' Catalog, Andorra Nurseries, Chestnut Hill, Philadelphia, Pa. 1908.
'Morris Dwarf' B. Wagenknecht in *The Boxwood Bulletin* 11(3):45.1972.

'Morris Midget' B. Wagenknecht in *The Boxwood Bulletin* 11(3):45.1972.
'Nana' Beissner, Schelle and Zabel, *Handbuch der Laubholz-Benennung* 283.1903.
'Nana Compacta' Catalog, Mayfair Nurseries, Bergenfield, N.J. 1954.
'National' D. Andberg in *The Boxwood Bulletin* 12(4):62.1973.
'Obcordata' Beissner, Schelle and Zabel, *Handbuch der Laubholz-Benennung* 283.1903.
'Obcordata Variegata' Anonymous in "List of Plants Introduced by Robert Fortune from Japan." *Gardners Chronicle* 735.1861.
'Rotundifolia' Beissner, Schelle and Zabel, *Handbuch der Laubholz-Benennung* 283.1903.
'Rotundifolia Glauca' Catalog, Charles Dietriche, Angers, France. 1892.
'Rotundifolia Pendula' Catalog, Andorra Nurseries, Chestnut Hill, Philadelphia, Pa. 1919.
'Rubra' T. Makino in *Botanical Magazine of Tokyo* 27:112.1913.
'Variegata' L. Dippel, *Handbuch der Laubholzkunde* 3.83.1893.

Buxus sinica var. *insularis* (Nakai) Hatusima in *Botanical Magazine of Tokyo* 36:63.1922.
'Cushion' B. Wagenknecht in *The Boxwood Bulletin* 7(1):1.1967.
'Staygreen' Catalog, John Vermeulen and Son, Neshanic Station, N.J. 1961.
'Tall Boy' B. Wagenknecht in *The Boxwood Bulletin* 7(1):1.1967.
'Tide Hill' D. Wyman in *Arnoldia* 17(11):64.1957.
'Winter Beauty' B. Wagenknecht in *The Boxwood Bulletin* 7(1):1.1967.
'Wintergreen' D. Wyman in *Arnoldia* 23(5):88.1963.

Buxus sempervirens Linnaeus, *Species Plantarum* 983.1753.
'Abilene' Inventory, Beal-Garfield Botanic Garden, East Lansing, Mich. 1960.
'Agram' Introduced by the USDA, Glenn Dale, Md. Spring 1959.
'Andersoni' A name applied to a number of plants grown from seed collected by Dr. Edgar S. Anderson in Macedonia. No precise application of the name seems possible.
'Angustifolia' P. Miller, *Gardener's Dictionary* ed. 8: Bux. no. 2. 1756.
'Angustifolia Variegata Punctulata' H. Baillon, *Monographie des Buxacées et des Stylocérées* 61.1859.
'Arborescens' P. Miller, *Gardener's Dictionary* ed. 8: Bux. no. 1. 1756.
'Arborescens Argentea' J. Loudon, *Arboretum et Fruticum Britannicum* III: 1333. 1838.
'Arborescens Aurea Acuminata' H. Baillon, *Monographie des Buxacées et des Stylocérées* 60.1859.
'Arborescens Aurea Punctulata' H. Baillon, *Monographie des Buxacées et des Stylocérées* 60.1859.
'Arborescens Decussata' *Kew Handlist of Trees and Shrubs* 269.1925.
'Argentea Nova' Catalog, V. Gauntlett, Chiddingfold, Surrey, England. 1930.
'Argenteo-variegata' R. Weston, *Botanicus Universalis* 1:31.1770.
'Aristocrat' J. Baldwin in *The Boxwood Bulletin* 6(2):23.1966.
'Aurea Pendula' *Kew Handlist of Trees and Shrubs* 131.1896.
'Aureo-variegata' R. Weston, *Botanicus Universalis* 1:31.1770.
'Belleville' R. Seibert in *Arnoldia* 23(9):116.1963.
'Blauer Heinz' Otto Markworth and Dr. Hans-Georg Preissel in *Deutsche Baumschule* p.516 Dec. 1987.
'Broman' Sheridan Nurseries, Toronto, Canada. Selected in 1936.
'Bullata' G. Kirchner in Petzold and Kirchner, *Arboretum Muscaviense* 194.1864.
'Butterworth' Catalog, Tingle Nurseries, Pittsfield, Md. 1958.
'Caucasica' Hort.ex K. Koch, *Dendrologie* v.2, pt. 2: 476. 1872.
'Christiansen' Catalog, Cary Brothers Nursery, Shrewsbury, Mass. 1957.
'Clembrook' E. Clements in *The Boxwood Bulletin* 8(2):20-22.1968.
'Cliffside' J. Baldwin in *The Boxwood Bulletin* 14(1):15.1974.

'Columnaris' Catalog, Visser's Nurseries, 132-9 Merrick Blvd., Springfield Gardens, Long Island, N.Y. 1960.
'Compacta' Catalog, Charles Dietriche, Angers, France. 1953.
'Conica' Catalog, Siebenthaler Nurseries, Dayton, Ohio. 136:10.1938.
'Crispa' Hort.ex K. Koch, *Dendrologie* v. 2, pt. 2: 476. 1872.
'Croni' Catalog, Monroe Nurseries, Monroe, Mich. 1955.
'Cucullata' Hort.ex K. Koch, *Dendrologie* v. 2, pt. 2: 476. 1872.
'Decussata' L. Dippel, *Handbuch der Laubholzkunde* 3:82.1893.
'Edgar Anderson' M. Gamble in *The Boxwood Bulletin* 13(2):26-28.1973.
'Elegans' L.H. Bailey, *Standard Cyclopedia*. 601.1914.
'Elegantissima' Hort.ex K. Koch, *Dendrologie* v. 2, pt. 2: 477. 1872.
'Fairview' Catalog, Eastern Shore Nurseries Inc., Easton, Md. 49. 1947.
'Fastigiata' F. Meyer in *Plant Explorations ARS* 34-9:91.1959.
'Glauca' G. Kirchner in Petzold and Kirchner, *Arboretum Muscaviense*. 194.1864.
'Glauca Marginata Aurea' Catalog, F. Delauney, Angers, France. 1910.
'Globosa' Catalog, Siebenthaler Nurseries, Dayton, Ohio. 136:10.1938.
'Graham Blandy' L. Batdorf in *The Boxwood Bulletin* 25(1):8.1985.
'Grandifolia' J. Mueller, Arg. in *De Candolle, Prodromus* 16(1):19.1869.
'Grand Rapids' Catalog, Light's Tree Company, Richland, Mich. 12:14.1948.
'Green Beauty' Catalog, Eastern Shore Nurseries, Easton, Md. 10.1964.
'Handsworthiensis' Fisher ex Henry in Elses and Henry, *Trees of Great Britain and Ireland* #7:1725.1913.
'Handsworthii' Hort.ex K. Koch, *Dendrologie* v.2, pt. 2: 476. 1872.
'Handsworthii Aurea' Catalog, Visser's Springfield Gardens, Merrick Rd., Springfield, Long Island, N.Y. 1945.
'Hardwickensis' Beissner, Schelle and Zabel, *Handbuch der Laubholz-Benennung* 283.1903.
'Hardy Michigan' Catalog, John Vermeulen and Son Inc., Neshanic Station, N.J. 1959.
'Harmony Grove' D. Wyman in *American Nurseryman* 107(7):57.1963.
'Hendersonii' Catalog, Lindley Nurseries, Greensboro, N.C. 1958.
'Heinrich Bruns' F. Meyer, New Cultivars of Woody Ornamentals from Europe in *Baileya* 9(4):129.1961.
'Henry Shaw' M. Gamble in *The Boxwood Bulletin* 25(2):43-47.1985.
'Hermann von Schrenk' M. Gamble in *The Boxwood Bulletin* 14(2):31-ibc. 1974.
'Heterophylla' V. Veillard in Duhamel, *Traité des Arbres et Arbrisseaux*, ed. augm. 1:82. 1835.
'Hood' M. Gamble in *The Boxwood Bulletin* 26(3):64-67.1987.
'Inglis' D. Wyman in *Arnoldia* 17(11):65.1957.
'Joe Gable' Catalog, Kingsville Nursery, Kingsville, Md. 1946.
'Joy' M. Gamble in *The Boxwood Bulletin* 24(1):12-13.1984.
'Latifolia Macrophylla' *Kew Handlist of Trees and Shrubs* 609.1902.
'Latifolia Maculata' *Kew Handlist of Trees and Shrubs* 131.1896.
'Latifolia Marginata' *Kew Handlist of Trees and Shrubs* 269.1925.
'Latifolia Nova' *Kew Handlist of Trees and Shrubs* 609.1902.
'Lynnhaven' Greenbrier Farms Inc., Norfolk, Va. 1922.
'Macrophylla' Beissner, Schelle and Zabel, *Handbuch der Laubholz-Benennung* 283.1903.
'Marginata' J. Loudon, *Arboretum et Fruticum Britannicum* III. 1333. 1838.
'Mary Gamble' J. Penhale in *The Boxwood Bulletin* 26(2):34-35.1986.
'Memorial' J. Baldwin in *The Boxwood Bulletin* 6(4):ibc. 1967.
'Minima' Beissner, Schelle and Zabel, *Handbuch der Laubholz-Benennung* 283.1903.
'Minima Glauca' Catalog, Charles Dietriche, Angers, France. 1892.
'Minor-aureo' R. Weston, *Botanicus Universalis* 1:31.1770.
'Mucronata' Hortul. ex H. Baillon, *Monographie des Buxacées et des Stylocérées* 62.1859.

'Myosotidifolia' *Kew Handlist of Trees and Shrubs* 131.1896.
'Myrtifolia' *Catalog of Trees and Shrubs*, Gordon, Dermer and Edmonds Pl. 6. 1782.
'Natchez' M. Gamble in *The Boxwood Bulletin* 26(3):62-63.1987.
'Newport Blue' Catalog, Boulevard Nurseries, Newport, R.I. 4.1941.
'Nigricans' P. Corbelli, *Dizionario di Floricultura* 232.1873.
'Nish' M. Gamble in *The Boxwood Bulletin* 14(4):61.1975.
'Northern Find' D. Wyman in *Arnoldia* 23(5):87-88.1963.
'Northern New York' Inventory, Beal-Garfield Botanic Garden, East Lansing, Mich. 1960.
'Northland' Registered by C.W. Stuart and Co., Newark, N.Y. 1949.
'Notata' R. Weston, *Botanicus Universalis* 1:31.1770.
'Oleaefolia' L.H. Bailey, *Standard Cyclopedia* 601.1914.
'Pendula' Catalog, Simon Louis 21.1869.
'Ponteyi' L. Dippel, *Handbuch der Laubholzkunde* 3:81.1893.
'Prostrata' W. Bean, *Trees and Shrubs Hardy in the British Isles* 1:278.1914.
'Pullman' W. Pullman in *The Boxwood Bulletin* 11(2):20-21.1971.
'Pyramidalis' Catalog, Simon Louis 21.1869.
'Pyramidalis Hardwickensis' *Kew Handlist of Trees and Shrubs* 269.1925.
'Pyramidalis Variegatis' Catalog, Baudriller Nursery, Angers, France. 1880.
'Rosmarinifolia' Hortul. ex H. Baillon, *Monographie des Buxacées et des Stylocérées* 62.1859.
'Rotundifolia' H. Baillon, *Monographie des Buxacées et des Stylocérées* 61.1859.
'Rotundifolia Aurea' L. Dippel, *Handbuch der Laubholzkunde* 3:82.1893.
'Rotundifolia Aureo-variegata' Beissner, Schelle and Zabel, *Handbuch der Laubholz-Benennung* 284.1903.
'Rotundifolia Maculata' F. Meyer in *Plant Explorations ARS* 34-9:113b.1959.
'Rotundifolia Minor' Beissner, Schelle and Zabel, *Handbuch der Laubholz-Benennung* 284.1903.
'Salicifolia' Hort.ex K. Koch, *Dendrologie* v.2, pt. 2: 476. 1872.
'Salicifolia Elata' Catalog, F. Delauney, Angers, France. 1896.
'Semi-elata' Catalog, Charles Dietriche, Angers, France. 1892.
'Semperaurea' B. Wagenknecht in *The Boxwood Bulletin* 7(1):1.1967.
'Serbian Blue' M. Gamble in *The Boxwood Bulletin* 14(4):61.1975.
'Ste. Genevieve' M. Gamble in *The Boxwood Bulletin* 11(1):1,15-16.1971.
'Subglobosa' Beissner, Schelle and Zabel, *Handbuch der Laubholz-Benennung* 283.1903.
'Suffruticosa' Linnaeus, *Species Plantarum* 983.1753. As *B. suffruticosa*.
'Suffruticosa Alba Marginata' Catalog, Brimfield Nurseries, Wethersfield, Conn. 1955.
'Suffruticosa Aurea' H. Baillon, *Monographie des Buxacées et des Stylocérées* 61.1859.
'Suffruticosa Aureo-marginata' Beissner, Schelle and Zabel, *Handbuch der Laubholz-Benennung* 284.1903.
'Suffruticosa Crispa' Beissner, Schelle and Zabel, *Handbuch der Laubholz-Benennung* 284.1903.
'Suffruticosa Glauca' Beissner, Schelle and Zabel, *Handbuch der Laubholz-Benennung* 284.1903.
'Suffruticosa Maculata' Beissner, Schelle and Zabel, *Handbuch der Laubholz-Benennung* 284.1903.
'Suffruticosa Variegata' R. Weston, *Botanicus Universalis* 1:31.1770.
'Tenuifolia' Hortul. ex H. Baillon, *Monographie des Buxacées et des Stylocérées* 61.1859.
'Thymifolia' Beissner, Schelle and Zabel, *Handbuch der Laubholz-Benennung* 283.1903.
'Thymifolia Variegata' *Journal of the Royal Horticultural Society* 18:82.1895.
'Undulifolia' *Kew Handlist of Trees and Shrubs* 270.1902.
'Vardar Valley' D. Wyman in *Arnoldia* 17(7):42-44.1957.

'Varifolia' Catalog, Kingsville Nurseries, Kingsville, Md. 1949.
'Welleri' Catalog, Weller Nursery Co., Holland, Mich. 1945.

Buxus x 'Green Gem' B. Wagenknecht in *The Boxwood Bulletin* 7(1):1.1967.

Buxus x 'Green Mountain' B. Wagenknecht in *The Boxwood Bulletin* 7(1):1.1967.

Buxus x 'Green Velvet' B. Wagenknecht in *The Boxwood Bulletin* 7(1):1.1967.

III. Excluded Cultivars

Buxus microphylla
'Kingsville Dwarf' D. Wyman in *American Nurseryman* 117(7):50.1963. = **'Compacta.'**

Buxus sempervirens
'Acuminata' *Journal of the Royal Horticultural Society* 18:86.1895. = *Buxus acuminata* J. Mueller, Arg. Buxaceae. *De Candolle, Prodromus* 16(1):15.1869.
'Albo-marginata' cited by D. Wyman in *American Nurseryman* 117(7):57.1963. = **'Argenteo-variegata.'**
'Angustifolia Variegata' J. Loudon, *Arboretum et Fruticum Britannicum* III. 1333. 1838. = **'Argenteo-variegata.'**
'Angustifolia Variegata Maculata' H. Baillon, *Monographie des Buxacées et des Stylocérées* 61.1859. = **'Argenteo-variegata.'**
'Arborescens Aurea' J. Loudon, *Arboretum et Fruticum Britannicum* III: 1333. 1838. = **'Aureo-variegata.'**
'Arborescens Aurea Maculata' H. Baillon, *Monographie des Buxacées et des Stylocérées* 60.1859. = **'Aureo-variegata.'**
'Arborescens Aurea Marginata' H. Baillon, *Monographie des Buxacées et des Stylocérées* 61.1859. = **'Marginata.'**
'Arborescens Gable' Catalog, Tingle Nurseries, Pittsfield, Md. 1963. = **'Joe Gable.'**
'Arborescens Longifolia' L. Dippel, *Handbuch der Laubholzkunde* 3:82.1893. = **'Angustifolia.'**
'Arborescens Marginata' J. Loudon, *Arboretum et Fruticum Britannicum* III: 1333. 1838. = **'Marginata.'**
'Arborescens Salicifolia' L. Dippel, *Handbuch der Laubholzkunde* 3:82.1893. = **'Salicifolia.'**
'Arborescens Tenuifolia' L. Dippel, *Handbuch der Laubholzkunde* 3:82.1893. = **'Angustifolia.'**
'Arborescens Thymifolia' H. Vogel, *Gartenwelt* 33:150.1929. = **'Thymifolia.'**
'Arborescens Variegata' Catalog, Andorra Nurseries, Chestnut Hill, Philadelphia, Pa. 1908. = **'Argenteo-variegata.'**
'Argentea' C. Ludwig, *Die Neuere Wilde Baumzucht.* 9.1783. = **'Argenteo-variegata.'**
'Argenteo-marginata' L. Dippel, *Handbuch der Laubholzkunde* 3:81.1893. = **'Argenteo-variegata.'**
'Aurea' J. Loudon, *Arboretum et Fruticum Britannicum* III. 1333. 1838. = **Aureo-variegata.'**
'Aurea Maculata' *Kew Handlist of Trees and Shrubs* 131.1896. = **'Aureo-variegata.'**
'Aurea Maculata Aurea' Inventory, Beal-Garfield Botanic Garden, East Lansing, Mich. 1960. = **'Aureo-variegata.'**
'Aurea Maculata Pendula' Inventory, Beal-Garfield Botanic Garden, East Lansing, Mich. 1960. = **'Aurea Pendula.'**
'Aureo-limbata' R. Weston, *Botanicus Universalis* 1:31.1770. = **'Marginata.'**
'Aureo-marginata' Beissner, Schelle and Zabel, *Handbuch der Laubholz-Benennung* 284.1903. = **'Marginata.'**
'Elata' L. Dippel, *Handbuch der Laubholzkunde* 3:82.1893. = **'Angustifolia.'**

'Elegantissima Variegata' Catalog, Charles Dietriche, Angers, France. 1892. = **'Elegantissima.'**
'Flavo-marginata' L. Dippel, *Handbuch der Laubholzkunde* 3:81.1893. = **'Marginata.'**
'Flavo-variegatis' Beissner, Schelle and Zabel, *Handbuch der Laubholz-Benennung* 284.1903. = **'Aureo-variegata.'**
'Fruticosa' Duhamel, *Arbres & Arbustes*, ed. 2, i.t. 24. 1801-19. = **'Suffruticosa.'**
'Fruticosa Foliis Variegata' F. Dietrich, *Vollstandiges Lexicon der Gartnerei und Botanik* 2:391.1802. = **'Suffruticosa Variegata.'**
'Gigantea' V. Veillard in Duhamel, *Traité des Arbres et Arbrisseaux*, ed. augm. 1:82. 1835. = *Buxus balearica* Willd.
'Golden' Commonly used in catalogs as a descriptive term = **'Aureo-variegata.'**
'Horizontalis' *Hilliers Manual of Trees and Shrubs* 1972. = **'Prostrata.'**
'Humilis' K. Koch, *Syn. Fl. Germ. Helv.* ed. 2, 8: 722. 1844. = **'Suffruticosa.'**
'Japonica Aurea' *Hilliers Manual of Trees and Shrubs* 1972. = **'Latifolia Maculata.'**
'Latifolia' Anonymous in *Annals of Horticulture* 2:541.1847. = **'Bullata.'**
'Latifolia Bullata' *Kew Handlist of Trees and Shrubs* 609.1902. = **'Bullata.'**
'Ledifolia' Beissner, Schelle and Zabel, *Handbuch der Laubholz-Benennung* 283.1903. = **'Salicifolia.'**
'Leptophylla' V. Veillard in Duhamel, *Traité des Arbres et Arbrisseaux*, ed. augm. 1:82. 1835. = **'Myrtifolia.'**
'Longifolia' G. Kirchner in Petzold and Kirchner, *Arboretum Muscaviense* 194.1864. = **'Angustifolia.'**
'Macrocarpa' Beissner, Schelle and Zabel, *Handbuch der Laubholz-Benennung* 283.1903. = **'Macrophylla.'**
'Macrophylla Glauca' Beissner, Schelle and Zabel, *Handbuch der Laubholz-Benennung* 283.1903. = **'Glauca.'**
'Macrophylla Rotundifolia' Beissner, Schelle and Zabel, *Handbuch der Laubholz-Benennung* 283.1903. = **'Rotundifolia.'**
'Maculatis' Beissner, Schelle and Zabel, *Handbuch der Laubholz-Benennung* 284.1903. = **'Argenteo-variegata.'**
'Myrtifolia Glauca' Beissner, Schelle and Zabel, *Handbuch der Laubholz-Benennung* 284.1903. = **'Suffruticosa Glauca.'**
'Nana' V. Veillard in Duhamel, *Traité des Arbres et Arbrisseaux*, ed. augm. 1: 83, 1835. = **'Suffruticosa.'**
'Navicularis' Catalog, Charles Dietriche, Angers, France. 1892. = **'Handsworthiensis.'**
'Oleaefolia Elegans' L.H. Bailey, *Standard Cyclopedia* 601.1914. = **'Elegans.'**
'Pyramidata' Inventory, Sanford Arboretum, Tenn. 1932. = **'Pyramidalis.'**
'Rosmarinifolia Crispa' Beissner, Schelle and Zabel, *Handbuch der Laubholz-Benennung* 284.1903. = **'Suffruticosa Crispa.'**
'Rosmarinifolia Fruticosa' P. Corbelli, *Dizionario di Floricultura* 1:232.1873. = **'Suffruticosa.'**
'Rosmarinifolia Major' H. Baillon, *Monographie des Buxacées et des Stylocérées* 62.1859. = **'Rosmarinifolia.'**
'Rosmarinifolia Minor' H. Baillon, *Monographie des Buxacées et des Stylocérées* 62.1859. = **'Suffruticosa.'**
'Suffruticosa Myrtifolia' Beissner, Schelle and Zabel, *Handbuche der Laubholz-Benennung* 284.1903. = **'Myrtifolia.'**
'Suffruticosa Nana' Catalog W.T. Smith Co., Geneva, N.Y. 1936. = **'Suffruticosa.'**
'Suffruticosa Navicularis' Beissner, Schelle and Zabel, *Handbuch der Laubholz-Benennung* 284.1903. = **'Handsworthiensis.'**
'Suffruticosa Rosmarinifolia' Beissner, Schelle and Zabel, *Handbuch der Laubholz-Benennung* 284.1903. = **'Rosmarinifolia.'**
'Suffruticosa Thymifolia' Catalog, Little and Ballantyne, Carlisle, England. 1928. = **'Thymifolia.'**
'Suffruticosa Variegata Maculata' H. Baillon, *Monographie des Buxacées et des Stylocérées* 61.1859. = **'Argenteo-variegata.'**

Public Gardens

United States

American Horticulture Society — Alexandria, Va.
Arnold Arboretum — Jamaica Plain, Mass.
Brooklyn Botanic Garden — Brooklyn, N.Y.
Buzzards Bay Garden Club — South Dartmouth, Mass.
Coker Arboretum — University of North Carolina, Chapel Hill, N.C.
College of William and Mary — Williamsburg, Va.
Colonial Williamsburg — Williamsburg, Va.
Dixon Gallery and Garden — Memphis, Tenn.
Eisenhower Center Library — Abilene, Kan.
George Landis Arboretum — Esperance, N.Y.
Longwood Gardens — Kennett Square, Pa.
Missouri Botanical Garden and Shaw Arboretum — St. Louis, Mo.
Morton Arboretum — Lisle, Ill.
Morris Arboretum — Philadelphia, Pa.
Mt. Vernon — Mt. Vernon, Va.
North Carolina State University Arboretum — Raleigh, N.C.
Old Westbury Gardens — Old Westbury, N.Y.
Preservation Society of Newport County — Newport, R.I.
Secrest Arboretum — Wooster, Ohio
State Arboretum of Virginia — Boyce, Va.
Tennessee Botanical Gardens and Fine Arts Center (also known as Cheekwood) — Nashville, Tenn.
U. S. National Arboretum — Washington, D.C.
University of Tennessee Arboretum — Oak Ridge, Tenn.
Washington Park Arboretum — Seattle, Wash.
Woodlawn Plantation — Alexandria, Va.

Europe

Hillier's Arboretum — Romsey, Hants., England
Kew Gardens — Royal Botanic Gardens, Kew, England
Royal Horticultural Society — Wisley, England
Oxford Botanic Gardens — Oxford, England
Royal Botanic Garden — Edinburgh, Scotland
Herrenhausen Garden — Hanover, Germany

Notes

Notes

Notes

Notes